Tropical Ecosystems in Australia

Responses to a Changing World

T0186731

Tropical Ecosystems in Australia

Responses to a Changing World

Dilwyn J. Griffiths

CRC Press
Taylor & Francis Group
Boca Raton London New York

CRC Press is an imprint of the
Taylor & Francis Group, an **informa** business

CRC Press
Taylor & Francis Group
6000 Broken Sound Parkway NW, Suite 300
Boca Raton, FL 33487-2742

International Standard Book Number-13: 978-0-367-36508-0 (Hardback)
International Standard Book Number-13: 978-0-367-34789-5 (Paperback)

Library of Congress Cataloging-in-Publication Data

Names: Griffiths, Dilwyn J., author.
Title: Tropical ecosystems in Australia : responses to a changing world / Dilwyn Griffiths.
Description: Boca Raton : Taylor & Francis, [2020] | Includes bibliographical references and index. | Summary: "Over the last century, the world has lived through changes more rapid than those experienced at any other time in human history, leading to pressing environmental problems and demands on the world's finite resources. Nowhere is this more evident than across the world's warm belt; a region likely to have the greatest problems and which is home to some of the world's most disadvantaged people. This book reviews aspects of the biology of tropical ecosystems of northern Australia, as they have been affected by climatic, social and land-use changes. Tropical Australia can be regarded as a microcosm of the world's tropics and as such, shares with other tropical regions many of the conflicts between various forms of development and environmental considerations"-- Provided by publisher.
Identifiers: LCCN 2019026394 (print) | LCCN 2019026395 (ebook) | ISBN 9780367347895 (paperback) | ISBN 9780367365080 (hardback) | ISBN 9780429328008 (ebook)
Subjects: LCSH: Rain forest ecology--Australia, Northern. | Rain forests--Australia, Northern.
Classification: LCC QH541.5.R27 G75 2020 (print) | LCC QH541.5.R27 (ebook) | DDC 577.340994--dc23
LC record available at https://lccn.loc.gov/2019026394
LC ebook record available at https://lccn.loc.gov/2019026395

Visit the Taylor & Francis Web site at
http://www.taylorandfrancis.com

and the CRC Press Web site at
http://www.crcpress.com

Printed and bound by CPI Group (UK) Ltd, Croydon, CR0 4YY

I dedicate this book to the memory of our much-loved daughter, Manon, who died on Saturday, 15th June 2019, after a long battle with cancer. An entomologist working in the Queensland Department of Agriculture, Fisheries and Forestry Group, and a devoted environmentalist and forest biologist, she will be fondly remembered by a wide circle of friends and colleagues within Queensland Science and by her work contacts in Southeast Asia.

Contents

Preface..xi
Acknowledgements xiii
Author xv
Introduction...xvii

Chapter 1   Tropical Forests 1

Wet Tropics World Heritage Area (WTWHA) 2
Biodiversity of Tropical Forest Flora 4
Rainforest Vascular Epiphytes 6
CAM Photosynthesis in Tropical Rainforest Epiphytes...................... 7
Climate Change Effects on Tropical Forests.. 8
Temperature Effects on Tropical Forests 13
Heat Injury in Tropical Plants 14
Temperature Effects on Photosynthetic Productivity.................... 14
Effect of CO_2 Concentration on Photosynthetic Productivity........... 15
Water-Use Efficiency in Tropical Forests... 17
Tropical Forests as Carbon Sinks... 18
Tropical Forest Tree Health.. 21

Chapter 2   Wetlands, Mangroves and Impoundments 23

Underground Water and Intermittent Rivers...................................... 23
Northern Wetlands 25
Billabongs, Farm Dams and Artificial Lakes 26
Farm Dams 26
Weirs and River Barrages.. 27
Artificial Lakes and Reservoirs.. 28
Mangroves... 29
Carbon Fluxes in Wetlands 31
Wetlands of Tropical Floodplains 32
Invasive Species 32
The Spread of Cane Toads Throughout Tropical Wetlands 33
Mimosa pigra in the Anson Bay Region 34
Invasive Tree, *Ziziphus* 34
Rubber Vine (*Cryptostegia grandiflora*) 35
Salvinia molesta, a Floating Aquatic Fern 36
Rubberbush (*Calotropis procera*)..................................... 36
Introduced Vermin 36

Chapter 3 Tropical Savannas 39

Vegetation of the Tropical Savannas 39
Savanna Soil Carbon 40
Fire Management.. 42
Savanna Burning and CO_2 Emissions 44
Desert Ecosystems 45

Chapter 4 Tropical Crops 47

The Australian Sugar Cane Industry... 48
  Green Cane Trash Blanketing (GCTB) 48
  Australian Sugar Cane and C-4 Photosynthesis............................. 49
  Sugar Cane under Elevated CO_2 Concentrations 50
  Ethanol from Sugar Cane 50
Other Tropical Broadacre Crops 51
  Cotton 52
  Rice... 53
  Sorghum 54
Legumes in Sustainable Agriculture... 55
  Legumes in Crop Rotations 56
Novel Crops 57
  Sandalwood 57
  Kenaf 58
  Agave 58
Tropical Fruit and Vegetables 59
  Protected Cropping in the Tropics.. 60
  The Banana Industry in North Queensland.................................... 60
  Mango (*Mangifera indica*)... 61
  Asparagus 62
Tropical Plantation Forestry.. 62
  Selective Logging... 63
  Plantation Forestry in Northern Australia..................................... 63
  Silviculture and Ecosystem Values... 65
Australia's Contribution to Forestry in South-East Asia.................. 66

Chapter 5 Rangelands and Tropical Pastures...................................... 69

Pasture Improvement... 69
The Northern Beef Cattle Industry 71
Rangelands and Climate Change 73
Regeneration of Abandoned Pastures... 74

Chapter 6 Marine Ecosystems 77

Oceanic Carbon Sequestration.. 77
Seagrass Beds of Coastal Areas.. 78

The Northern Fisheries Industry 79
    The Northern Prawn Fishery 80
    Barramundi in Northern Australia 80
    Mud Crab Fisheries 80
    Other Recreational Fisheries 81
    Indigenous Fisheries 81
Tropical Aquaculture 81
    Barramundi 82
    Prawn Culture 83
    Giant Clams 83
Reef Ecosystems 84
    Coral Bleaching 84
    Carbon Fluxes in Corals under Elevated CO_2 and Temperature 85
    Bleaching in Giant Clams 86
    Crown-of-Thorns Starfish 86
    Ningaloo Marine Park, Western Australia 87
Great Barrier Reef and Tourism 88

Chapter 7   Mining and Mine-Site Rehabilitation 91

The Ranger Uranium Mine at Kakadu 91
Mining at Coronation Hill 93
Mining in the Pilbara 94
The Rum Jungle Mine 94
Mining in the Mount Isa Region 95
Bauxite Mining at Weipa 98
Kidston Gold Mine 99
Open Cuts and Big Pits 100
Coal-Seam Gas Production 101
Legislative Aspects of Mine-Site Rehabilitation 102

Chapter 8   National Parks and Conservation 105

National Parks of Tropical Queensland 105
National Parks of the North of Western Australia 107
    The Kimberleys 107
    The Pilbara 108
National Parks of the Northern Territory 109
Joint Management of National Parks 109

Chapter 9   Management Issues 111

Rural Industries of the North 112
Indigenous Australians and Tropical Land Management 113
Wild Rivers Act 114

                Buffalo and Wild Pig in Arnhem Land 114
                Domiculture Practices of Cape York Peninsula 115

Chapter 10 Conclusions 117

                Postscript 123

References 125

Index 151

Preface

The world has, over the last hundred years or so, lived through changes more rapid than those experienced at any other time in human history. It has been a period of significant and continuing increases in world population, rapid technical advances, increasing industrialisation and the associated impacts on the environment. These changes have presented us with challenges that seem almost beyond our capacity to meet. The environmental problems and our demands on the world's finite resources may, arguably, be the most pressing of all. Nowhere is this more evident than in the world's warm belt, a region likely to have the greatest problems, yet one that is, in many parts of the world, home to some of the most disadvantaged people.

This book deals specifically with the tropical north of Australia, which, up until recent times and despite its being blessed by a wide range of natural resources, has always been sparsely populated compared with the rest of the country and certainly in comparison with other tropical regions of the world. The region can be regarded as a microcosm of the tropical world and its problems, yet with many biological, climatological, geological and socio-economic differences that set it apart from the tropics generally. The main focus of this book is on the region's ecosystems and how they have been affected by the changes in land use brought about by the post-colonial and post-federation development of northern Australia. More recent developments include an expansion of crop production and a reassessment of the north's potential as a source of food and other resources for the region. Particular attention will be given to ecosystem responses to climate change and to the challenges to ecosystem management that will have to be faced if we are to halt (and perhaps reverse) the environmental decline that has been all-too-evident in some areas over recent decades.

Ecosystems under particular threat include the Great Barrier Reef off the east coast of north Queensland, the coastal mangroves across the north, the wet tropics rainforest remnants and the vast hinterland of arid to semi-arid savanna country – many of them already showing signs of environmental degradation but with a clear capacity to respond to appropriate restoration measures. Other, more resilient ecosystems will also be described, particularly those able to adapt to environmentally friendly forms of development balancing profitability with sustainability.

Acknowledgements

The information contained in this book is drawn from the published work of a number of professional biologists (and those from related disciplines) who have studied various aspects of the natural history of what was, up until some 50 to 100 years ago, a little-known region of Australia. Many of these people were work colleagues who generously shared with me their knowledge of the region. Many more I have known only through their contribution to the ever-growing scientific literature relating to the tropical north. Some I have been fortunate to have as travelling companions throughout the region and as fellow members, from time to time, of relevant committees, working groups and professional societies. They are all warmly acknowledged and their input duly recorded in the reference list. I apologise to those whose work I may have missed or, for various reasons, omitted to include in this review. I take full responsibility for any errors and omissions.

It is a pleasure to record that the work described here owes much to the inputs available from a range of government-supported research and development projects established and funded under the Australian government's Cooperative Research Centres (CRC) program. Commencing in 1990 (and continuing since that time), the program has fostered collaborative research partnerships involving industry, research organisations and universities. Centres of particular relevance to the subject matter of this book include CRCs on Tropical Rainforest Ecology and Management, Tropical Savanna Management, Sustainable Sugar Production and Management of the Great Barrier Reef World Heritage Area. Other government-funded research projects making significant contributions to specific topics relating to the ecology and sustainable development of northern ecosystems include the Tropical Rivers Inventory and Assessment Project 2005–2008, Tropical Rivers and Coastal Knowledge Research 2007–2011, and the CSIRO Northern Australia Sustainable Yield Flagship Project "Water for a Healthy Country", all highlighting the key role of the region's freshwater resources in the sustainable development and preservation of northern Australian ecosystems.

Although this book deals specifically with tropical Australia, there may be principles and problems described that are applicable to the tropics generally. To assist those readers not familiar with the geography of northern Australia, I have included some maps showing the major localities and bioregions referred to in the text. My grandson, Cai Hughes', assistance in preparing these maps is warmly acknowledged.

I thank those colleagues (and anonymous referees) who read and commented on earlier drafts of the manuscript. Finally, I thank the University of Queensland for allowing me access to its library facilities as a guest user.

Acknowledgements

Author

Dilwyn J. Griffiths, born in 1932, was brought up in a rural farming community in West Wales. A graduate of the University of Wales, UK, his post-graduate (PhD) training specialised in studies of the growth and metabolism of microalgae. After a short period as a high school teacher in Cardiff, he was appointed Assistant Lecturer (later Lecturer) in Plant Physiology at the University College of North Wales, Bangor, UK. In 1967, he was appointed Senior Lecturer at the newly established La Trobe University, Melbourne, Australia, and, in 1974, was appointed to the Chair of Botany at James Cook University, Townsville, Australia. On retirement in 1997, he was appointed Professor Emeritus at the same university, but now lives in Brisbane. He is the author or co-author (with colleagues or as supervisor of post-graduate research programs) of publications dealing with various aspects of the biology of marine and freshwater microalgae and with studies of the freshwater resources of the tropical north of Australia. Since his retirement, he has continued to write reviews and has taken an active interest in current developments in his area of biological research. He is the author of three recently published books: *Microalgal Cell Cycles* (2010), *Microalgae and Man* (2013) and *Freshwater Resources of the Tropical North of Australia: A Hydrobiological Perspectiv*e (2016). He is a Fellow of the Royal Society of Biology, UK, and a long-term member and supporter of the Royal Society of Queensland, Australia. Dilwyn and his wife, Elen, who is also a graduate of the University of Wales, have two daughters and a son and eight grandchildren – all living in Australia.

Introduction

In terrestrial Australia, the tropics (an area of ca. 3.1×10^6 km^2 lying to the north of the Tropic of Capricorn) comprises a variety of natural and developed habitats. They include remnant rainforest areas along the north-east coastal region of Queensland and along the north coast of Australia as far as the Kimberly/Pilbara region of Western Australia, extensive coastal mangrove and wetland areas, vast areas of inland tropical savanna merging into extremely arid desert country and large areas under various forms of agricultural, industrial and urban development. The offshore continental shelf supports commercial and recreational fishing, extensive coral reef ecosystems such as the Great Barrier Reef off the east coast of Queensland and a large fringing reef (Ningaloo Reef) off the west coast of Western Australia.

All these ecosystems have been exposed to massive changes over the period dating from the initial invasion and colonisation of Australia by European (and other) settlers and the subsequent opening up of the north to development over the last 150 or so years. These changes have been more rapid than any previously experienced, none more so than those associated with the observed changes in atmospheric CO_2 (and their presumed consequent climatic changes), all with implications for the economy of the region and for the health and well-being of its population.

An assessment of the vulnerability of Australian tropical biomes to these changes has now become a major concern, particularly in view of the clear indications that they are occurring at rates well in excess of any hitherto experienced. They are driven by the unprecedented expansion of various forms of agricultural, industrial and social developments into the tropical north as well as by global climatic changes more dramatic than any experienced since the most massive extinction event of the Cretaceous–Tertiary (K–T) boundary (ca. 66×10^6 years BP) (Steffen et al., 2009).

Geological evidence suggests that for at least 800,000 years before the start of the 20th century, global atmospheric CO_2 concentrations probably never exceeded 300 ppm. By 2011, however, concentrations of 392 ppm had been recorded, with predictions that values in excess of 800 ppm may be expected by 2100 (Luthi et al., 2008). It is now generally accepted that a significant portion of this increase may be due to anthropological effects, particularly related to industrialisation and the increasing use of fossil fuels – oil, coal and gas. Increasing concentrations of atmospheric CO_2 (and other gases with similar atmospheric effects) have been implicated in what has been described as the "greenhouse effect", namely a surface warming which predictive models have suggested may, in the absence of remedial action, have the potential to lead to a global warming of between 1.8 and 3.6°C by the end of the 21st century (Meehl et al., 2007). Other major contributors to global warming are methane (CH_4) emissions from livestock (enteric fermentation of ruminants) and livestock manure, rice growing (especially rice paddy cultivation), solid waste landfills (allowing anaerobic decay of organic matter) and coal, oil and natural gas mining. Natural CH_4 sources include wetlands, oceans and anaerobic sediments (Milich, 1999). CH_4, although a potent greenhouse gas, is relatively short-lived in the atmosphere where it undergoes a hydroxyl radical reaction process.

One of the major consequences of global warming, already evident from climate data over recent decades, is the greater frequency of extreme climate events. What were previously 1 in 100-year extreme events have been predicted to trend, in many areas, towards 1 in 10-year events. The tropical north of Australia would be particularly sensitive to such changes because its productive capacity is already severely constrained by the highly fragile and nutrient-poor soils, its extreme climatic conditions and its seasonally highly variable freshwater availability.

For vast areas of Australia's tropical region, the effects of climate change are likely to be compounded by the fact that the biota and ecosystems are already under pressure from other drivers of change such as habitat fragmentation, spread of invasive species, reduced availability of freshwater sources and increased pollution. As has been the case in sub-tropical and temperate Australia (and elsewhere in the world), there is already a trend in the tropical north towards the release of previously low-productivity farming lands for rehabilitation as natural or semi-natural ecosystems.

The climate changes now being experienced pose unprecedented challenges to conservation, primarily because the predicted rates of change will exceed anything experienced over historical times. It has been estimated that a 2–3°C increase in global temperatures above pre-industrial levels will, for example, place 20–30% of species at risk of extinction by 2100 (Thomas et al., 2004). Particularly at risk will be those species already designated as threatened or endangered. More prolonged dry seasons may increase the frequency, intensity and extent of wild fires, with some forests transitioning from forest to savanna or grassland.

The major emphasis of this review will be directed towards effects upon the vegetation and other primary producers because they, through their photosynthetic processes, are most immediately affected by the climatic and environmental changes, with flow-on effects throughout the ecosystem. They are also the most immediately affected by increasing atmospheric CO_2 concentrations, and through various forms of carbon sequestration have the potential to act as a major natural sink of carbon and hence a potential means of mitigating the continuing build-up of atmospheric CO_2. The environmental health and sustainability of the various ecosystems described and their ability to survive and/or adapt to the changing conditions and pressures will depend largely on these primary producers and on our ability to manage wisely their vital role in their semi-natural or developed habitats.

1   Tropical Forests

In tropical Australia (Figure 1.1), the rainforest flora contains a unique association of species, descended from ancestors which became established during the Late Cretaceous/Early Tertiary and were later modified by subsequent climatic changes and by the northward drift of the Australian tectonic plate (Morley, 2000). The flora is mostly autochthonous (with a characteristically Australian sclerophyll element) as dictated by the presumed wide separation of the Australian and south-east Asian tectonic plates before the latter part of the Tertiary (Hill, 1994). These factors have resulted in the present-day unique character of Australia's tropical forest flora with its high concentration of primitive angiosperms (Barlow & Hyland, 1988), and a variety of arborescent and climbing palms, ferns and epiphytes including many endemic species. Many genera are limited to a single representative. They are all the survivors of the extensive drying that occurred during the middle and late Miocene and the more recent Pliocene (Morley, 2000).

Their species richness, with no clearly dominant species, has been interpreted as reflecting the major crises they have experienced since their first appearance ($\sim 75 \times 10^6$ years BP) including the meteorite impacts of the Cretaceous, the Eocene cooling, the Late Neogene expansion of the polar ice caps as well as the tectonic movements of the Late Tertiary. They have many features that distinguish them from other tropical forests of the world such as those of the neotropics (South America, the West Indies, Central America south of Mexico), tropical Africa and even the more adjacent Asia-Pacific region.

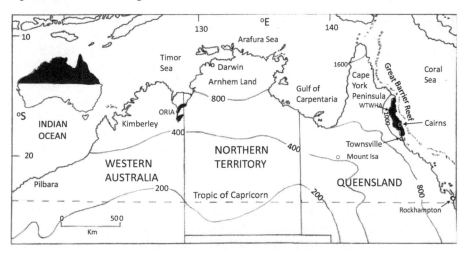

FIGURE 1.1   The tropical north of Australia, showing major locations and bioregions referred to in the text. The annual rainfall isohyets (mm) are estimations based on Australian Bureau of Meteorology data averaged over the period 2000 to 2010. WTWHA, Wet Tropics World Heritage Area; ORIA, Ord River Irrigation Area.

The so-called wet tropics areas occupy regions experiencing annual rain-falls of between 1,000 and 1,600 mm. The principal plant families repre-sented in today's tropical rain forests of northern Australia are the Myrtaceae, Lauraceae, Elaeocarpaceae, Rutaceae and Proteaceae. It has been suggested (e.g. Burbidge, 1960) that the present composition of the Australian flora is pri-marily due to climate selection both within the region and from biotypes avail-able as a result of migration (by communities rather than by chance dispersal of individuals). Most of these tree families are endemic to Australia and only a small number of species extend beyond continental Australia although they may belong to (or have allies in) various regions of south-east Asia (Francis, 1981). It is deduced from geological evidence that rainforest covered much of northern Australia during the early to middle Miocene ($23–15 \times 10^6$ years BP) but has since that time become restricted to the present-day remnants remaining after the extensive drying of the continent (Long et al., 2002). The fossil record shows little evidence for a post-Miocene influx of Asian plants into Australian rainforests (Ashton, 2003).

These forests have in the past contributed timber products and cabinet woods of great economic value, but a large area, extending as a narrow coastal strip (~100 km wide and 400 km in length) along the eastern side of Cape York is now designated as the Wet Tropics World Heritage Area and protected from further exploitation. Despite its comparatively small size (8,940 km^2) relative to other tropical rainforest regions of the world, it is important not only for its conserva-tion value as a remnant of what was a unique and widespread ecological biome but also for its status as a carefully managed and protected area relatively free of the economic and social pressures that threaten rainforests elsewhere (Primack & Corlett, 2005).

WET TROPICS WORLD HERITAGE AREA (WTWHA)

Australia became a signatory of the World Heritage Convention in 1974 and by 2007 there were 17 Australian properties listed. One of them, the Wet Tropics of Queensland World Heritage Area, was listed in 1988. The listing was a recognition of its outstanding natural values including its importance to an understanding of the evolution of the world's ecosystems as influenced by geological, geomorphic and physiographic features. It was also a response to current reports (e.g. Tracey, 1982) that approximately one-quarter of the total area of rainforest of northern Queensland had been cleared during the 20th century, mainly for intensive production, grazing and agriculture.

The WTWHA (Figure 1.2) is managed by the Wet Tropics Management Authority, answerable to both the Australian Government and the Queensland State Government but working in partnership with other bodies responsible for conser-vation, research, community engagement and Aboriginal involvement. Day-to-day maintenance of the area is carried out by the appropriate land managers includ-ing the State Environmental Protection Agency, Department of Natural Resources and various local government authorities (Australian Wet Tropics World Heritage Authority, 2007–2008).

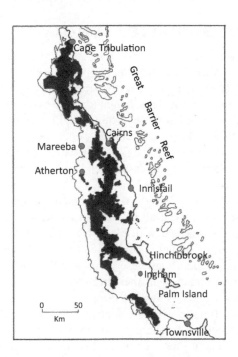

FIGURE 1.2 Wet Tropics World Heritage Area (shaded) extending from 15°26′ S in the north to just north of Townsville (19°16′ S) in the south. (A Directory of Important Wetlands in Australia, 2001, 3rd edition, as published in the Annual Report and State of the Wet Tropics 2010–2011 and reproduced here under the Creative Commons Australia Attribution Licence [CC BY], http://creativecommons.org/licenses/by/au.)

Management of the WTWHA recognises different zones established according to ecosystem integrity, its remoteness from disturbance, the intended physical and social setting, and management purpose. The least pristine zone defined as "compromised forest" is that upon which developed facilities (such as visitor facilities) are allowed. This zone might also include land upon which logging has in the past occurred (before 1988, when commercial logging ceased). Reclassification of such land would depend on the extent of its recovery. The WTWHA has been certified as a tourist destination (Esparon et al., 2013) and is recognised as such by a range of interested parties including tourist operators, environmentalists and volunteer groups.

The combination of high topographic, edaphic and climatic variability within the WTWHA is reflected in a very complex mosaic of different forest types. This has allowed development of models to describe different types of forest growth in terms of the prevailing physiological and physical processes. Two such models, 3-PG (a forest-growth model based on site and climate data) and 3-PGS (driven primarily by satellite data of PAR (photosynthetically active radiation), have been proposed by Nightingale et al. (2008). These models, calibrated and validated against field measurements, contribute to a better understanding of the carbon dynamics of Australia's tropical rainforest bioregions.

BIODIVERSITY OF TROPICAL FOREST FLORA

A detailed floristic study of a 25 ha monitoring plot of rainforest situated west of Atherton (17°07′ S, 145°37′ E) in the Wet Tropics bioregion of northern Australia (Sattler & Williams, 1999) was conducted by Bradford et al. (2014). Their initial census of plants having diameters at breast height (dbh) 10 cm recorded 208 species, 128 genera from 53 families with Lauracaea, Rutaceae, Proteaceae and Elaeocarpaceae as dominant. About 83% of the plants recorded were endemic to Australia and 45.2% were endemic to the Wet Tropics bioregion. The data collected provided a baseline description of the rainforest flora against which changes in floristic and stand structure could be measured.

The vegetation of this sample plot was described as complex mesophyll vine forest on granite and meta-sediment alluvium with moderately low soil fertility. The climate of the region is seasonal with 61% of annual rainfall occurring between January and March (mean annual rainfall ~1597 mm). The floristic and structural analysis of the recorded data was described as reflecting the unique nature of Australian wet tropical forests with their high species endemism and affiliations with both Indo-Malayan and Gondwanan flora.

One of the main climatic events affecting the forests of the wet tropics is that due to cyclones, whose frequency and severity are expected to increase as part of anticipated global climatic changes. Studies carried out at the Australian Canopy Crane Research Facility at Cape Tribulation (Figure 1.2) (Stork et al., 2007) have compared the floristic structure of a 0.95 ha lowland tropical forest plot over a 5-year period following the passage of Tropical Cyclone Roma through the area in February 1999 (Laidlow et al., 2007). A survey of the crane plot in 2005 examined 680 stems (≥10 cm dbh) of 82 tree species and showed a post-cyclone increase of 30% in number of stems and a 16% increase in number of species falling into this category. This was interpreted to indicate that the crane site had undergone an active period of post-cyclone recovery suggesting a high degree of temporal stability within the plot despite suffering frequent catastrophic disturbance.

Comparison of the floristic make-up of the crane plot with that of other lowland plots nearby and with mid-elevation plots elsewhere in the north Queensland region showed that the Cape Tribulation site (with *Cleistanthus myrianthus*, *Alstonia scholaris*, *Myristica insipida*, *Normanbya normanbyi* and *Rockinghamia augustifolia* as the most abundant species) supports a floristically unique community. There was also considerable variation between rainforest communities across the north Queensland region reflecting differences in local topography, environment and disturbance history.

Today's wet tropics forest remnants of north-eastern Australia are threatened by a crisis that has the potential to be as destructive as any of those experienced in the past. In addition to the immediate economic consequences of the loss of a range of plant and animal products, the loss of biodiversity, already evident in temperate regions to the south, would impact particularly upon the special role of tropical forests as the primary gene pool from which most temperate plant taxa are thought to have evolved (Bews, 1927). The fossil record shows that the rates and levels of forest destruction now feared, especially if compounded by the effects of climate change, would take many millions of years for full recovery (Morley, 2000).

Forest diversity can be expressed in terms of taxonomic composition, functional and spatial features (including canopy stratification) and the physiological characteristics of the constituent taxa. All of these will influence how the forest responds to climatic or other changes. Fast-growing tree species, for example, may respond more readily to elevated CO_2 concentrations than slow-growing species. Equally, species growing under limited nutrient supply may respond differently to CO_2 enrichment compared with species under unlimited nutrient supply. For these reasons, investigations of the responses to climate change are best applied under conditions of minimal interference (i.e. growth conditions that are as natural as possible).

The high diversity of rainforest flora, often with the higher taxonomic groups represented by one or a few species, brings particular threats to survival, especially in times of change. *Idiospermum australiense*, a rainforest canopy tree restricted to a few small populations in north-east Queensland, is the only southern hemisphere representative of the Calycanthaceae. The flowers are protogynous, with some populations being andromonoecious whilst others are hermaphrodite. This species is insect pollinated during its 10–16 day floral life span (Worboys & Jackes, 2005), and although pollen production is large, it is likely that spatial and temporal separation of male and female functions may result in greater vulnerability to climatic change.

Species occupying a specialist niche in the rainforest ecosystem may be particularly vulnerable. Examples are provided by those cases where the life history of the plant and that of an associated animal have become closely linked through a process of co-evolution, as appears to be the case with the rare rainforest tree *Ryparosa* sp. (Achariaceae) and its associated fruit-eating, ground dwelling avian cassowary (*Casuarius casuarius johnsonii*). Studies at the rainforest of the Daintree National Park of far north Queensland by Webber and Woodrow (2004) have shown that whilst the young fruit of the Ryparosa have a high concentration of cyanogens, the ripe fruit (consumed by the cassowary) have negligible cyanogen content. The authors further report that passage through the cassowary digestive system improved the germination rate of the *Ryparosa* seed from 4% to 92% with a much reduced susceptibility of the seed to fruit-fly larvae. Clearly, the intricate balance between the plant and its associated frugivore is crucial to the survival of both partners and the ecosystem.

Observations from other tropical forests of the world (e.g. Laurance et al., 2004) have noted that while tree mortality, recruitment rates and growth have increased over recent decades, there has been a decline in the dominance or density of the slower-growing trees, including species occupying the sub-canopy. Such compositional changes might be expected to impact upon the capacity for carbon storage (see later) and, of course, upon the other forest biota. The extent to which these changes are linked to climate changes is the subject of considerable current speculation.

Any significant loss of biodiversity is known to affect the integrity and functioning of forest ecosystems (Scherer-Lorenzen et al., 2005) confirming observations from early forest research that the yield from mixed forests (on good soils) can, on average, be up to 10–20% (or even up to 50%) higher than that from monoculture in the same area (Kenk, 1992). It is generally accepted that loss of diversity (e.g. from forest clearing) is often accompanied by other environmental problems such as degradation of waterways. In recognition of this, a number of community groups in northern Australia have received government assistance to engage in re-planting and re-establishment programs to restore biodiversity (Erskine, 2002).

RAINFOREST VASCULAR EPIPHYTES

The vascular epiphytes of the north Australian forests are dominated by ferns and orchids rather than the bromeliads that are such a prominent feature of the epiphytic flora of neotropical forests (Crayn et al., 2015). A study of the distribution of epiphytic ferns and orchids in relation to light and moisture availability has recently been conducted by Sanger and Kirkpatrick (2016). The study was extended to a lower montane rainforest in the Mount Lewis region (16°30′ S, 145°12′ E) in a north Queensland Wildlife Sanctuary managed by the Australian Wildlife Conservancy and the Mount Lewis National Park. Elevations ranged from 800 to 1,180 m with samples taken and measurements made at different heights on the host trees from forest floor to canopy.

A strong partitioning of different epiphyte taxonomic groups was noted relative to the light and moisture gradients. Orchids tended to occur in sites that have the better access to light but quite high mean daily maximum vapour pressure deficits (VPD 0.43 kPa); ferns occurred in more shaded sites with lower VPD (0.28 kPa), i.e. more humid locations. Neither of the major groups of epiphytes has any direct connection with the ground; they rely on the host plants for support but they are not parasitic on the host. Their distribution will therefore be strongly influenced by microclimate factors that create niche locations. The major limiting factor would appear to be moisture availability. A secondary factor would be the high light preference of the orchids and the low light preference of the ferns (and hemi-epiphytes). Thus, the orchids occur in the more exposed habitats (e.g. outer canopy) while the ferns occur in the lower, moister habitats, usually in the inner canopy of host trees. These preferences are reflected in some of the morphological characteristics such as the pseudobulbs, thickened leaves and specialised roots of the orchids – all features likely to be beneficial under drought conditions. Many ferns also have thickened leaves together with a tendency to form "baskets" which accumulate litter (and hence moisture) around the roots.

The growth and survival of the epiphytic fern *Asplenium nidus* (bird's nest fern) in the canopy of a lowland tropical rainforest in the Daintree region of north Queensland (Freiberg & Turton, 2007) were strongly affected by the severity and frequency of the recurring periods of drought experienced in that region. Four continuous weeks without rain, for example, completely dried out the fern's accumulated humus. More prolonged dry periods killed the roots that attached the healthy fern to the host tree causing the epiphyte to fall to the ground. Dry periods of more than 8 weeks resulted in the death of fern plants even in the most protected locations such as host branch forks. Estimations of the age of *A. nidus* plants (based on a number of morphological criteria) showed that, across a 1 ha plot, the oldest specimens dated from just after the most recent period of prolonged drought.

Any increase in the frequency of periods of prolonged drought, such as has been predicted might occur over future decades, would surely impact on the survival of *A. nidus* to the point that the fern might come to be restricted to the lower strata of the forest with accompanying effects on canopy fauna and other epiphytes such as *Ophioglossum, Psilotum, Schellolepis* and *Lycopodium*.

CAM PHOTOSYNTHESIS IN TROPICAL RAINFOREST EPIPHYTES

Many, but not all, epiphytic plants are assisted in their capacity for survival under conditions of low moisture availability by their adoption of a specialised form of photosynthetic CO_2 fixation (the CAM mechanism), which has been shown to be particularly suited to such conditions. The crassulacean acid metabolism (CAM) photosynthetic carbon dioxide fixation mechanism is so-called because it was first discovered in succulent plants belonging to the family Crassulaceae. It has since been shown to occur in a wide range of other angiosperm families including, among others, the Cactaceae, Orchidaceae, Bromeliaceae, Liliaceae and Euphorbiaceae. CAM plants usually grow in arid regions or where water is otherwise not easily available such as salt marshes or epiphytic locations (as described earlier), where the CAM mechanism allows the plants to take in CO_2 (through their open stomatal apertures) during the night (when loss of water by transpiration would be minimal). The CO_2 reacts with a 3C acceptor (phosphoenolpyruvate) to form the 4C acids oxaloacetic, malic and other 4C acids, stored in the vacuoles (hence the sour taste detectable in succulent plant leaves at night and early morning). The energy driving this reaction comes from starch glycolysis. During daylight (with the stomata now closed to prevent loss of water) the accumulated 4C acids are decarboxylated releasing CO_2 to be internally photosynthetically fixed by the reactions of the Calvin cycle to form photosynthetic products (including the storage starch needed to drive the nighttime intake and fixation of CO_2 (HCO_3^-)) (Black & Osmond, 2003).

For tropical epiphytes, therefore, the CAM photosynthetic mechanism would be well suited to their canopy locations where they have the benefit of exposure to high light intensities, reduced photoinhibition and the water-conserving attributes afforded by the CAM photosynthetic mechanism (Winter et al., 1983). The last of these would be particularly important since epiphytes obtain their water supply either directly from rainfall or from the humid air of the canopy. On a worldwide basis, well over 50% of epiphytic species have been estimated to undergo CAM photosynthesis compared with only ~10% of vascular plants generally (Lüttge, 1997). In the Australian tropics, however, the epiphytic flora is dominated by epiphytic and lithophytic fern species, which, generally, are thought to have only a very low incidence of CAM expression.

More detailed studies with two species of Australian tropical epiphytic ferns (*Microsorium punctatum* and *Polypodium crassifolium*) and a lithophytic fern (*Platycerium veitchii*) showed that *Polypodium* and *Platycerium* did indeed exhibit weak CAM characteristics (Holtum & Winter, 1999). Although there was no net nocturnal CO_2 uptake, the presence of CAM was inferred from nocturnal increases in titratable acidity, a reduction in the rates of net CO_2 evolution during the first half of the dark period and the presence of a CAM-like decrease in net CO_2 uptake during the early light period. *M. punctatum* also displayed some CAM characteristics with a net uptake of CO_2 during the first half of the dark period (and an increase in titratable acidity) and a pronounced reduction in net CO_2 uptake during the early light period.

Another criterion that has been used to distinguish plants having a CAM-type carbon fixation mechanism is the stable carbon isotope ratio ($\delta^{13}C$). Carbon atoms in organic matter exist in the form of two stable isotopes, ^{12}C and ^{13}C, and their relative amounts (R) in any organic sample ($^{13}C/^{12}C$) are expressed as $\delta^{13}C$, which represents, in parts per thousand, the difference between the sample and a reference material (Peterson & Fry, 1987):

$$\delta^{13}C = \left[\left(R_{sample} / R_{standard} \right) - 1 \right] \times 10^3$$

- δ^{13} C values for C-3 plants (most trees, shrubs and temperate grasses) are approximately –28‰.
- δ^{13} C values for C-4 plants (mainly tropical and salt-land grasses) are approximately –12‰.

Applying the $\delta^{13}C$ test to the aforementioned tropical epiphytic species gave values –22.6 to –25.9‰ (clearly more like C-3–type ratios; see Winter et al., 1983), which were interpreted by Holtum and Winter (1999) as indicating that carbon isotope ratios may not, by themselves, be sufficient to indicate the presence of CAM, if it is weak. A number of environmental factors such as water availability, irradiance, altitude and reutilization of respired CO_2 can all affect the $\delta^{13}C$ value in plant tissue, sometimes at the ecosystem level. It would appear, then, that plants of such ancient lineages as ferns may have at least some of the range of CAM CO_2 exchange patterns that occur in higher plants. CAM plants are considered to be highly adaptable and the expression of CAM is under both developmental control and environmental influences. Little wonder, then, that CAM plants such as cacti, bromeliads and sedums and various succulents have been shown to have good environmental survival abilities.

CLIMATE CHANGE EFFECTS ON TROPICAL FORESTS

As a preliminary to their study of how the forests of Australia's wet tropics are likely to be affected by climate change, Hilbert et al. (2001) adopted the criteria first described by Webb (1959) and Tracey and Webb (1975) to define a number of different structural forest types ranging from dry open woodland at one extreme to humid lowland forest at the other. The range of criteria used for the classification included canopy heights, degree of canopy closure, presence or absence of vines, dominance of sclerophyllous species and a number of other diagnostic features such as leaf size of canopy trees, abundance of epiphytes, lianas and plank buttressing.

According to this classification, the humid tropics region, extending from Cooktown (15°28′ S, 145°15′ E) in the north to just north of Townsville (19°16′ S, 146° E) in the south and in patches from the north of Cape York Peninsula across the

FIGURE 1.3 Vegetation of tropical Australia (as described by Wood, 1950) and reproduced here with permission from CSIRO Publishing, Australia. A – monsoonal woodland; B – savanna woodland and brigalow scrub; C – sclerophyllous grass steppe; D – Desert sclerophyllous grassland; E – open savanna; F – mulga scrub; G – grassy scrub; shaded areas – tropical rain forest. The isohyets labelled 9–9 indicate the limit of an annual growing season of >9 months duration; those labelled 5–5 indicate the limit of an annual growing season of 5 to 9 months.

north coast (Figure 1.3), could be viewed as a mosaic of environments each suited to a particular forest type. Climate change would therefore have the potential to affect the spatial distribution of this mosaic, leading to stresses and eventually to altered forest distributions. Using an Artificial Neural Network analysis method, Hilbert et al. (2001) were able to predict how sensitive the different forest types might be to different forms of climate change. They tested 15 different structural forest types and 10 different climate scenarios that might be expected over the next 50–100 years.

Among the most interesting findings were:

1. Increased precipitation might be expected to favour some rainforest types, whereas decreased rainfall would be likely to increase the area suitable for forests dominated by sclerophyllous genera (e.g. *Eucalyptus* and *Allocasuarina*).
2. The area of lowland mesophyll vine forest environments was predicted to increase with warming, whereas the response of upland complex notophyll vine forests would depend on precipitation.
3. Highland rainforest environments (simple notophyll and simple microphyll vine fern forests and thickets) were predicted to decrease by 50% with only a 1°C warming. Several forest types were predicted to become highly stressed by a 1°C warming and most would be sensitive to any change in rainfall.

4. Most forest types were predicted to experience climates that would be most
   suited to a different forest type so that the propensity for ecological change
   in the region would thus be high, leading to the likelihood of significant
   shifts in the extent and spatial distribution of forests. The strongest effects
   of climate change were predicted to be experienced at boundaries between
   forest types or between rainforest and open woodland.

Climate change effects on forest ecosystems will be most critically felt at the forest
canopy, the interface between the atmosphere and the biosphere. Models of the most
likely responses have to consider temperature, light, rainfall and humidity, storm
severity and CO_2 concentrations. They will also have to consider the complexity of
the ecosystems with their mix of species and the interactions between them through
the different trophic levels (Stork et al., 2007).

The range of effects on tropical forests now attributed to climate change include
direct effects that induce physiological stress as well as interactions with other cli-
mate-mediated processes such as insect attack or outbreaks of wildfire. Data from
worldwide studies reviewed by Allen et al. (2010) indicate that some of the world's
forested ecosystems have already experienced higher rates of tree mortality with
consequent loss of the capacity for carbon sequestration and associated atmospheric
feedback (see later).

In the rainforests of north-eastern Australia, the annual rainfall is high but most
of the rain is confined to the summer wet season. This is a particular characteristic of
the semi-arid tropics that they do not share with the more predictable climate of the
monsoonal forests further north. The seasonally dry rainforest communities suffer dry
periods of varying frequency and duration, and usually comprise a mix of phenological groups with islands of rainforest deciduous woody taxa within the more dominant
sclerophyllous vegetation. Multi-year droughts have been reported to trigger widespread
mortality of eucalypts and *Corymbia* (Fensham & Fairfax, 2007) and tree death in
Acacia woodland (Fensham & Fairfax, 2005) indicating a particular vulnerability to
climate change and to El Niño climatic events.

The aforementioned results may be regarded as extreme cases of the earlier
findings of Unwin and Kriedemann (1990) who monitored the drought tolerance
of different rainforest tree species along a 160 km transect extending from coastal
evergreen forest near Cairns, through rainforest on the Atherton Tableland to semi-
deciduous vine forest south-west of Mount Garnet (west of the Great Dividing
Range) (Figure 1.4).

The transect represented an annual rainfall gradient from 2,800 mm at the coastal
site, through 1,400 mm on the tableland to 760 mm at the inland site. The water
relations and tree growth of two tree species (*Acacia aulocarpa* – an evergreen)
and (*Melia azedarach* var. *australasica* – white cedar) were compared at three sites
along the transect. *Acacia* trees were clearly able to endure persistent low moisture
status with increasing seasonal drought stress from coast to inland with dry-season
water potentials ranging from –2.1 MPa to –6.4 MPa. For *Melia* trees, the mean
minimum leaf water potential values ranged from –1.7 MPa to –2.3 MPa across all
sites and seasons but, because of their deciduous habit, were able to avoid the impact
of drought. Annual patterns of growth (measured as increases in stem cross sectional

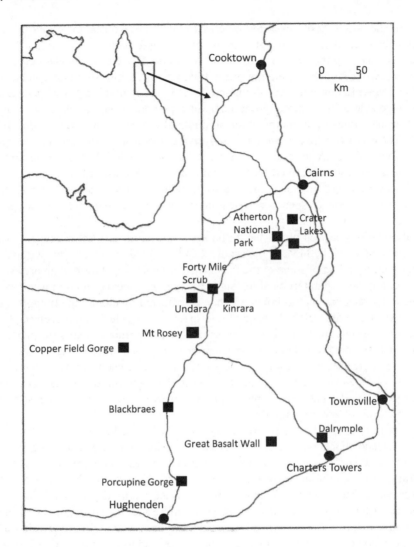

FIGURE 1.4 North-east coastal Queensland from Cooktown to Townsville showing the major towns (shaded circles), connecting highways and location of national parks (shaded squares) of the basalt provinces. (Redrawn with permission of the Geological Society of Australia [Queensland Division] from Willmott, W., 2009, *Rocks and Landscapes of the National Parks of North Queensland.*)

area) also varied according to species and site. For *Acacia*, the measured annual growth increment was 54.1 cm^2 year^{-1} (coastal site) and 56.9 cm^2 year (tableland site) but only 3.1 cm^2 year^{-1} at the inland site. It was concluded that seasonal drought was an important selection criterion in the establishment of rainforest species and one that would be strongly influenced by climate change.

Studies carried out at the Many Peaks Range (19°11′ S, 145°45′ E) rainforest of north Queensland compared the diurnal and seasonal pattern of leaf gas exchange and water relations of tree species of contrasting leaf phenology (Choat et al., 2006).

Two of the species, *Brachychiton australis* and *Cochlosspermum gillivraei*, were deciduous while the other two, *Alphitonia excelsa* and *Austromyrtus bidwillii*, were evergreen. During the transition from wet season to dry season, the deciduous species suffered a 70–90% reduction of the total canopy, while stomatal conductance (g_s) and assimilation rate (A) were much lower in the remaining leaves. These species were able to maintain daytime leaf water potentials (ψ_L) at close to or above wet season values by a combination of stomatal regulation and reduction of leaf area. The evergreen species were less sensitive to the onset of drought and g_s values were not significantly lowered during the transitional period. In the dry season, the evergreen species maintained their canopies despite increasing water stress. These findings were interpreted as being consistent with the view (Eamus & Prior, 2001) that tropical deciduous species compensate for shorter life spans with higher photosynthetic capacity.

Observations carried out at the 42 ha Daintree Rainforest Observatory at Cape Tribulation (16°06′ S, 145°26′ E) (Figure 1.2) have made a major contribution to a better understanding of some of the factors affecting the vegetation of an area which is typical of the lowland tropical forests of the region. Special attention was given to plant water relations and to how they may be affected by changing climatic conditions (Apgaua et al., 2015). A range of tree species were selected representing broad functional groups including one palm species and a number of mature phase and early successional dicotyledonous species. In situ rates of volumetric sap flow and velocities were measured using the heat ratio method. Results showed that all the species tested showed very little variation in sap flow between species and concluded that they were all capable of adopting appropriate hydraulic strategies in response to varying levels of water availability.

More frequent (and more severe) droughts and increased aridity, as are predicted under climate change scenarios, may result in various forms of hydraulic stress in different tree species, leading to loss of biomass or even increased mortality. The hydraulic architecture of various tree species would be particularly relevant to their performance or even survival, especially in the seasonally dry lowland rainforests of north-eastern Australia. The resilience of rainforest trees to drought conditions or precipitation extremes will depend, in particular, on the strategies available for the transportation of water and on their capacity to respond to water deficits. Among the range of contributing factors will be the rates of volumetric sap flow as affected by stem and leaf characteristics. Another factor that might influence drought resistance would be wood structure and its effect upon xylem conductivity or resistance to xylem embolism.

Concern that drought-induced tree mortality might offset any carbon sink benefits arising from old-growth or re-growth forests has highlighted the need for reliable models linking such effects to tree physiology. This has drawn attention to the plant vascular hydraulic transport system as the likely major trigger of tree mortality. Anderegg et al. (2016) analysed species mortality rates across 475 species from 33 study sites worldwide and found that mortality was best predicted by what they described as a low hydraulic safety margin, defined as the difference between typical minimum xylem water potential and that causing xylem dysfunction (e.g. xylem vulnerability to embolism).

Cyclones are the most extreme weather conditions encountered in the tropical upland rainforests of north Queensland. Gleason et al. (2008) assessed the structure and productivity of a range of forest plots before and after exposure to Cyclone Larry (March, 2006). They estimated the damage sustained by a range of tree species growing in nutrient-poor schist soils and those growing in relatively nutrient-rich basalt soils. Most damage (in terms of leaf area reduction and stem breakage) was sustained by trees growing on the basalt soils. Trees growing on nutrient-limited soils were, for some reason, less susceptible to the damaging effects of high winds.

TEMPERATURE EFFECTS ON TROPICAL FORESTS

Current expectations that global temperatures will continue their present trend into the future has focussed attention on how these temperature changes may affect inputs and outputs of carbon. Global models have projected that mean temperature increases in the tropics will be either in line with the global mean (1.7–3.9°C) or will be greater (1.8–5.0°C). This suggests that perhaps tropical biomes should receive particular attention in any ecosystem-scale climate change research. Meta-analysis of the effects of temperature on carbon storage and fluxes across a broad range of tropical forest sites found that total net primary production (NPP), litter production, tree growth and below-ground carbon allocation all increase with increasing mean annual temperature (MAT) and with increasing temperature to precipitation ratio (Wood et al., 2012). Decomposition of soil carbon and turnover time also increased with increasing MAT indicating that atmospheric carbon uptake could to some extent be offset by increased soil carbon loss due to warming.

Leaves of tropical forest trees are already exposed to very high temperatures at times of full solar radiation and may often experience severe heat injury. Such injury would be compounded under the extended dry seasons that would accompany any increase in the frequency of the El Niño years of the Southern Oscillation climatic cycle (Meehl et al., 2007), as also would any large-scale deforestation (with consequent increase in atmospheric CO_2 concentrations and reduced surface evaporation) (Cramer et al., 2004). Conditions that stimulate stomatal closure (such as elevated CO_2 concentrations) and consequent reduced cooling from transpiration, would produce leaf temperatures considerably higher than air temperatures. High temperatures have often been shown to display interactive effects with elevated CO_2 concentrations. Eucalypt seedlings grown under elevated temperature and CO_2 have been shown to have enhanced growth but reduced leaf tissue nitrogen (Ghannoum et al., 2010). But high temperature also increases plant water use such that the improved water-use efficiency (WUE) (induced by high CO_2 at ambient temperatures) is negated (Sherwin et al., 2012). The reduction in tissue nitrogen at high CO_2 and temperature was attributed to changes in root nitrogen uptake by mass flow due to altered transpiration rates at elevated CO_2 concerntrations and temperature.

Even those species showing no obvious sign of heat stress may often be exposed to temperatures that are supra-optimal for photosynthesis. As a number of reviews (e.g. Norby & Luo, 2004; Wood et al., 2012) have highlighted, increasing atmospheric temperature is only one of a number of interacting factors that influence how tropical forests respond to global warming. Effects upon ecosystem-scale processes and upon a range of forest biota such as endemic vertebrates (beyond the scope of this review) are predicted to result in large-scale shifts in species ranges and mismatch with seasons and food availability, with the possibility of extinctions (Williams et al., 2003; Couper et al., 2005).

HEAT INJURY IN TROPICAL PLANTS

Heat injury in plants can be assessed by noting the onset of increases in fluorescence of leaf chlorophyll. This has been tested for a range of different plant species comprising tropical, temperate and alpine plants, by slowly warming dark-adapted leaves at a rate of 1°C per minute (Smillie & Nott, 1979). For all the species tested, chlorophyll fluorescence began to increase between 30 and 40°C and reached a peak at around 50°C, but both the initial rise and the peak occurred at higher temperatures in the tropical plants.

A more detailed study of the high-temperature tolerance of two tropical tree species (*Ficus insipida*, an early successional species, and *Virola sebifera*, a late successional species) was carried out by Krause et al. (2010). This study measured the heat-induced increase of the initial fluorescence emission (F_o) as well as the decrease in the ratio of variable to maximum fluorescence (F_v/F_m). The latter measure, determined 24 h after heat treatment, was observed to be the best indicator of permanent tissue damage. For both species, the limit of leaf thermotolerance was between 50 and 53°C, only a few degrees more than the peak leaf temperature measured in situ. There was no marked seasonal variation of leaf thermotolerance and only a marginal increase of thermotolerance of plants grown under elevated temperatures. Irreversible leaf damage at 51–53°C was also observed in heat-stress experiments using intact potted seedlings although the plants recovered and developed new leaves during post-culture. The apparent limited capacity of these tropical tree species to acclimate to increased temperatures would be a major factor influencing their survival under minimal global warming.

When the response of seedlings of *Eucalyptus saligna* to ambient temperature (a) and a + 4°C was tested at three levels of atmospheric CO_2 concentration (280, 400 and 640 μLL^{-1}), it was shown that, at ambient temperature, water-use efficiency improved with increasing CO_2 concentration (Sherwin et al., 2012). But at the highest CO_2 concentration tested, because of the greater leaf area and high transpiration rates, water use was also higher. With a combination of a higher temperature and high CO_2 concentration, water-use efficiency was little changed.

TEMPERATURE EFFECTS ON PHOTOSYNTHETIC PRODUCTIVITY

Photosynthetic productivity of rainforest trees is modified by the prevailing conditions, with different species varying widely in their optimum temperature for photosynthesis and in their high-temperature compensation points. Trials with a number of Australian rainforest trees were set up to examine the relative ability of tropical species (compared with temperate species) to acclimate to changing temperatures (Cunningham & Read, 2003). Net photosynthesis of eight tree species was measured in leaves of seedlings maintained under constant day:night temperature regimes

(22:14°C) or fluctuating (27–19°C:17–9°C) regimes and the net photosynthesis measured under a constant temperature after 14 days acclimation. It was found that, in temperate species, at least 80% of their maximum net photosynthesis was achieved over a large span of acclimation temperatures. Tropical species, on the other hand, had a much lower acclimation potential, consistent with the larger seasonal day-to-day temperature variations of temperate climates. The apparent low capacity of the tropical species for temperature adjustment might therefore be a significant factor affecting their survival under extreme temperatures.

It has often been observed (e.g. Cernusak et al., 2013) that the net photosynthesis of tropical forest trees may approach zero at midday due to stomatal closure. Slot et al. (2016) determined the temperature response of leaf photosynthesis and respiration for seedlings of three tropical tree species from forests in Panama. It was noted that early successional species such as *Ficus insipida* and *Ochroma pyrimidale* had high temperature optima for photosynthesis and higher rates of photosynthesis at optimum temperature compared with the late successional species tested, *Calophyllum longifolium.* Canopy leaves of *F. insipida,* despite their apparent greater temperature tolerance did, however, suffer marked decrease of stomatal conductance at the higher temperatures with complete suppression of photosynthesis at ~45°C. It was suggested that, in addition to the effects on stomatal conductance, the higher temperatures may also stimulate photorespiration or, alternatively, may cause deactivation of rubisco or perhaps may affect membrane properties.

EFFECT OF CO_2 CONCENTRATION ON PHOTOSYNTHETIC PRODUCTIVITY

Because tropical forest trees are overwhelmingly C-3 species, and hence thought to have evolved in ancient geological times of very high atmospheric CO_2 concentrations (some 4 or 5 times present-day levels), atmospheric CO_2 concentrations of around 300 ppm are clearly severely limiting to photosynthetic carbon fixation. For them, therefore, any significant increase in atmospheric CO_2 concentrations would have the potential to stimulate forest photosynthetic productivity – the so-called fertiliser effect.

Plants growing close to the floor in tropical forests are known to be exposed to elevated concentrations of CO_2 compared with plants occupying above-canopy sites. Measurements carried out at 10 cm from the forest floor of Panamanian lowland and montane tropical forest sites, for example, yielded CO_2 concentrations of 387 and 423 LCO_2 L^{-1} for wet and dry seasons respectively (Holtum & Winter, 2001). Such modest levels of CO_2 enrichment (well below those required to saturate photosynthesis) were nevertheless deemed to be sufficient to increase the rate of CO_2 uptake in some species (such as the understorey shrub *Piper cordulatum* and the late successional tree species *Virola surinamensis*) relative to those attainable at above-canopy CO_2 concentrations. At the low photosynthetic photon flux densities normally encountered close to the forest floor, such elevated CO_2 concentrations (due to respiratory CO_2 diffusion from soil organisms and from decomposition of leaf litter) would increase photosynthetic performance and hence the survival of understorey seedlings (Winter & Virgo, 1998). Similar photosynthetic enhancement might therefore be a general feature of forest trees exposed to the elevated CO_2 concentrations predicted as part of global climate change.

Free-air CO_2 enrichment (FACE) experiments have been widely used to test for the response of natural vegetation or crops to elevated atmospheric CO_2 conditions (e.g. Hendrey et al., 1999). In such experiments, plants growing under otherwise natural field conditions are exposed to ~370 ppm CO_2 (control) or enhanced CO_2 (600 ppm). Examples of such experiments carried out on selected Australian tropical plant species (Holtum & Winter, 2003) have shown that shoots and leaves exposed to such enhanced CO_2 concentrations displayed increased rates of net CO_2 uptake (28% for *Tectona grandis*, teak, and 52% for *Pseudobombax septenatum*, barrigon). Interestingly, when the elevated CO_2 was applied as oscillating cycles of 20 seconds (a common feature of many trial experiments) with amplitude of ~170 µl CO_2 l^{-1} (mean 600) the stimulatory effect of the elevated CO_2 was much less (19% for teak and 36% for barrigon). Further studies with *T. grandis* in which the enhanced CO_2 was provided as 40 second oscillating cycles gave similar results in that the fertilising effect of the enhanced CO_2 was considerably less than that obtained if it was supplied continuously. It was concluded that FACE trials, in which the high CO_2 is invariably subject to short-term oscillations, tend to underestimate the potential fertilising effect of the higher CO_2 concentrations.

The fertilising effect of atmospheric CO_2 enhancement is strongly influenced by other factors. It is greatly reduced, for example, under low nutrient conditions (Kimball, 2011). It has been suggested that C-3 plants growing under enhanced CO_2 concentrations experience a change in carbon flux and in the pattern of nitrogen allocation such that allocation of nitrogen to key photosynthetic enzymes is optimised with respect to carbon flux (Woodrow, 1994). This has been interpreted to occur through a feedback mechanism that smooths out any imbalance within the photosynthetic system caused by a rise in atmospheric CO_2.

Modelling studies have indicated that the stimulation of photosynthesis under elevated CO_2 concentrations may be more marked in the tropics than it is at higher latitudes. This may be because photorespiration (a process that normally represents a loss of CO_2) is suppressed at higher temperatures (Cernusak et al., 2013). It has been observed that as leaf temperature increases, the specificity of the enzyme Rubisco for fixing CO_2 rather than O_2 also decreases. The solubility of CO_2 also decreases relative to that of O_2 at higher temperatures. Such effects are, however, likely to vary greatly between different forest vegetation types.

Short-term CO_2 enhancement experiments have generally shown a steep positive response of CO_2 assimilation to higher atmospheric CO_2 concentrations (Winter, 1979). Over the long term, however, assimilation rates and plant growth are clearly moderated by various feedback responses and constraints. Franks et al. (2013) have described the response of plants to increasing (or decreasing) atmospheric CO_2 as being characterised by an adaptive feedback that tends to maintain a relative constant gradient for CO_2 diffusion into the leaf. This gradient can be described by the relationship

$$1 - Ci / Ca$$

where Ca is the atmospheric CO_2 concentration and Ci is the concentration of CO_2 within the leaf. It has been suggested that the relationship is kept constant by predictable adjustments to stomatal anatomy and chloroplast biochemistry. For any given set of environmental conditions, the rate of exchange of CO_2 and water vapour at the leaf surface is governed largely by the diffusive conductance of the epidermis as determined by the density, pore size and dynamic properties of the stomata.

Atmospheric CO_2 is only one of the many environmental signals to which stomata respond. In addition to the short-term sensitivity of stomatal pore size, there are more permanent changes of stomatal density which translate to shifts in maximum leaf conductance observable over longer timescales. Franks et al. (2013) have conducted an evaluation of the extent to which observations of stomatal sensitivity measured over short-term experiments adequately describe long-term responses that may be modified by resource constraints, developmental processes or adaptation.

WATER-USE EFFICIENCY IN TROPICAL FORESTS

The effect of elevated CO_2 concentrations on water-use efficiency, well documented in field and glasshouse experiments, is likely to be of particular relevance to the response to climate change in plants subjected to prolonged dry periods. One of the most frequently observed effects of exposure of plant leaves to higher CO_2 concentrations is a reduction in stomatal openings with consequent reduction of evaporative loss. Conversely, exposure of leaves to lower than normal concentrations of CO_2 will cause the stomata to open. The magnitude of the response to enhanced CO_2 varies greatly from species to species but would be of particular advantage to plants exposed to conditions of limited water availability such as those of the Australian wet–dry tropics.

Seedlings of the Australian tropical forest tree species *T. grandis* kept in the dark, released 4.8% less respiratory CO_2 when exposed to high CO_2 (Holtum & Winter, 2003). Earlier work with seedlings of another Australian tropical species *Maranthes corymbosa*, showed that exposure to CO_2 enrichment resulted in a short-term decline of stomatal conductance as a result of decreased stomatal aperture (Berryman et al., 1994). It was also reported that for trees of *Eucalyptus tetrodonta* grown under elevated CO_2 concentration, the slope of the response of stomatal conductance to temperature was lower and the equilibrium stomatal conductance at saturating light was also lower.

The mechanisms involved in the relationship between elevated CO_2 and stomatal conductance are complex, but the site for sensing the level of CO_2 is thought to be located within the guard cells. Whatever the mechanism, there is evidence from both laboratory and field trials that the yield of some crop plants, for example, increases by 10–45% when exposed to higher CO_2 concentrations. Such CO_2 effects are, however, very much reduced in plants grown in poor soils with limiting nutrient supply. Both nitrogen and phosphorus availability have been reported to constrain the response of Australian tropical forest trees to elevated CO_2, as is the case for forest trees generally. Thus, seedlings of *Pinus taeda* exposed for 2 years to CO_2 enrichment with or without nutrient additions showed increased rates of photosynthesis only when they also received supplemental nitrogen (Tissue et al., 1993). It was further noted that, at the elevated CO_2 concentrations, proportionately less nitrogen was allocated to rubisco (the carboxylating enzyme) and more to the light reaction component of photosynthesis. This suggests that the magnitude of the response to a high CO_2 environment would be very much governed by soil fertility.

In their review of studies of the response of a range of young tropical tree species to CO_2 augmentation, Holtum and Winter (2010) noted only small increases in above-ground carbon content and even less for older trees. They conclude that the response of forest vegetation to increased atmospheric CO_2 concentrations is more a matter of water-use efficiency rather than one of photosynthetic carbon fixation. It has been suggested (Cernusak et al., 2013) that tropical forest canopy leaves, in particular, may already be close to their high temperature threshold and that this moderates any photosynthetic stimulation. CO_2 augmentation was, however, noted to result in decreased transpiration and increased soil-water content (in both young and older trees), indicating that the major effect of CO_2 augmentation on the above-ground biomass of rainforest trees may be through an effect on soil-water pools and on fluxes through the soil–plant–atmosphere continuum.

Studies at a seasonally dry riparian tropical forest on the Atherton Tableland (17°22′ S, 145°32′ E) (Drake & Franks, 2003) (Figure 1.4) showed seasonal variation in both the origin of the water used by the forest plant community and in the water-use efficiency of the trees, as indicated by hydraulic conductivity (K_s). The trees were mostly moisture-dependent species representative of the formerly more extensive rainforest that covered the region before the widespread clearing for agricultural development. δO^{18} analysis of water from the soil, from groundwater and from tree xylem vessels was undertaken during the wet season and during the dry season as well as measurement of hydraulic xylem conductivity. In three of the five species tested, there was a significant loss of hydraulic conductivity during the dry season, which was positively correlated with differences between wet season and dry season midday leaf water potential. Plants less susceptible to loss of conductivity clearly had greater control over transpiration and were therefore more water-use efficient. It was concluded that the differences between the more and the less susceptible plants were largely related to water-use strategies and xylem hydraulic properties than to any partitioning of water resources. If this applied generally, it would indicate a strong causal link between stomatal control of transpiration rate and susceptibility to xylem embolism.

TROPICAL FORESTS AS CARBON SINKS

It has been estimated (e.g. Lal, 2004) that, on a global basis, more than 30% of the increase in CO_2 emissions since the industrial revolution may be due to changes in land-use practices such as deforestation, biomass burning, conversion of natural to agricultural systems, drainage of wetlands and soil cultivation. Earlier analyses carried out by Houghton et al. (1987) have indicated that deforestation in the world's tropics is a major contributor to the net release of carbon to the atmosphere. The rate of land clearing in northern Australia, despite some reduction over recent decades, is still a major component of land management. Protecting forested land does, however, come at a cost to the landholder and to the state, measurable as the return foregone by not converting to agriculture.

Other costs are those associated with installation, management and maintenance practices to optimise the capacity of the forest for carbon capture. Agricultural land stores very little carbon in its above-ground biomass, whereas standing tropical forests may contain 300–400 tons of CO_2 per ha (Kindermann et al., 2008). Land clearing is often accompanied by burning of vegetation on site, leading to immediate emission of carbon or more long-term emission following decomposition over time. Improved forest management practices such as increasing the forest rotation age have the potential to produce measurable increases in the carbon stock in the landscape (Sohngen & Brown, 2008). Re-planting forests rather than relying on natural revegetation after clearing or forest fire or cyclone damage is also known to increase the overall amount of carbon on site.

Tropical forests, worldwide, have been estimated to contain approximately one half of the total carbon stored in the world's terrestrial biosphere and to account for about one third of global terrestrial productivity (Pan et al., 2011). Australia's tropical forests, like those elsewhere (Flint & Richards, 1991), would have been subject to past human disturbance and by natural events such as fires or extreme weather and would probably not be in anything close to their primary state. Their biota would not therefore be in a steady state with respect to carbon gain (photosynthesis) or loss (respiration), and it has been suggested that although the biomass of these tropical forest may represent a major store of carbon, the amount is probably less, per unit area, than could potentially be stored. The residence time (t_w) of woody biomass in tropical forest ecosystems is highly variable and can range from 23 to 129 years (Galbraith et al., 2013), with a median value of 50 years. In both neotropic and palaeotropic forests t_w was highest in heavily weathered soils with low fertility or soils whose physical condition does not favour woody biomass degradation. Forests on fertile soils were reported to fix slightly more carbon and to allocate slightly more of that carbon towards growth while most of the net primary productivity of such plots was allocated to the production of fine roots which, according to the authors (Daughty et al., 2014) are unlikely to greatly enhance soil carbon stocks.

All forests leach C into the soil profile and eventually into rivers. Within a given area of forest, photosynthesis removes more C from the atmosphere than is returned via respiration of biota or loss through leaching or by fire. The difference is that stored in above- and below-ground biomass, in necromass (dead organic matter) and soil organic matter. C storage in forest soils can, however, be enhanced through conversion to charcoal (which is more resistant to decomposition) (Brown et al., 1992a).

Any assessment of the significance of tropical forest ecosystems as a source or sink of carbon requires quantitative estimations of the above-ground biomass of different forest types based on data drawn from long-term forest inventory plots. Such assessments, based on measurements of trunk diameter, wood density and height have been made for secondary and old-growth forests; for dry, moist and wet forests; for lowland and montane forests; and for mangrove forests. It has been reported (e.g. Chave et al., 2005) that trunk diameter is generally the best predictor (with some overestimations) of above-ground biomass followed by wood specific gravity, tree height and forest type. Refinements of the methods used to estimate above-ground biomass (and above-ground carbon content) have recently been devised (Réjou-Méchain et al., 2017) using statistical packages incorporating data for the different variables mentioned earlier. Although photosynthetic CO_2 sequestration in tropical ecosystems may be high, rates of respiration (such as that accompanying litter degradation) may also be high resulting in a reduced net productivity.

Typically, mature forest species have higher density woods than early colonising species (Whitmore & Silva, 1990) and have been estimated to accumulate C at the rate of 1–2 mg ha^{-1} year^{-1} (Brown et al., 1992a), although stands recovering from acute disturbance can display increased levels of C accumulation. In previously logged forests in the Malaysian peninsula, for example, recovery (in the form of ingrowth of new trees entering the minimum diameter class of 10 cm) contributes significantly to carbon storage, although it is the larger trees (breast height diameter >70 cm) that represent the major long-term carbon sinks (Brown et al., 1992b) despite their generally low density (< 5 per ha). Such large trees can continue to accumulate C for many decades or even centuries. Fast-growing plantation species have been estimated to accumulate C at rates as high as 15 mg C ha^{-1} year^{-1} (Lugo et al., 1988). In late secondary and mature tropical forests, necromass (coarse woody debris) may account for between 10 and 40% of the above-ground biomass with the potential to act as a long-term (up to 100 years) carbon sink.

Soil organic matter, another long-term location for stored carbon (Schlesinger, 1990) is particularly significant in forests ecosystems. Reafforestation of agricultural land has been shown to lead to recovery of the capacity for carbon storage, especially under moist conditions (Brown et al., 1992a). In their review of the current status of carbon sequestration in Australian forests, Christopher et al. (2012) reported that the potential for such sequestration has scarcely been seriously considered for forests nationwide, far less for those in the tropics. The scope for a substantial contribution from tropical forests is, however, considerable given the extensive acreages involved (especially of savanna woodland) and the potential for expanded afforestation or reforestation.

Carbon forestry in Australia, as described in the Carbon Farming Initiative Act, 2011 (Australian Government, 2011), is now recognised (e.g. Grace & Basse, 2012) as a component of the range of land-based initiatives that can be used to mitigate carbon emissions as well as contributing to various other environmental benefits. Its application has been criticised, however, as having the potential to displace land otherwise suitable for food production, although land subjected to afforestation/reafforestation can, of course, contribute to other environmental co-benefits such as controlled harvesting for plantation timbers and timber products.

Biological carbon sequestration rates for some plantation species commonly used in Australian tropical forests have been estimated using simulation models of above- and below-ground biomass (Grace & Basse, 2012). The total area required to offset 1,000 t of CO_2 (over a 30-year period) was found to range from 1 ha (for *Eucalyptus cloeziana* growing under high-productivity conditions in coastal north Queensland) to 45 ha (for *Auraucaria cunninghamii* growing in low-productivity conditions in inland central Queensland).

In the seasonally dry tropics (of northern Australia), evergreen, deciduous and semi- and brevi-deciduous trees very often co-occur. The different phenological groups show adaptations to the prevailing soil and atmospheric conditions particularly with regard to water availability. Cost-benefit analysis (Eamus, 1999) has suggested that leaves of deciduous species have high rates of metabolism but are not long lasting, while those of evergreen species are long lasting but with lower rates of metabolism. The sequestration of carbon dioxide by forests is a function of the balance between the substantial uptake of CO_2 by vegetation and CO_2 release through soil respiration. There is also a minor contribution of CH_4 by methanotrophic bacteria and of N_2O as a by-product of nitrification and denitrification reactions in the soil. It has been estimated (Blais et al., 2005) that, worldwide, the carbon sink is greatest in tropical forests (-632mg C m^{-2} d^{-1}). Some of this sink can however be offset by N_2O emissions from the soil.

One of the key processes involved in sequestration of carbon in tropical rainforests is the deposition of photosynthetic products in the sapwood either as xylem tissue or as other components of the woody tissue such as lignin and other inclusions. It has been observed that tropical timbers characteristically display large annual variation in tree ring width (including missing rings) reflecting large temporal variation in diameter growth. Schippers et al. (2015), using a method that compared simulated sapwood growth with 30-year tree ring records in Thailand, confirmed that sapwood growth in the tropical species *Toona ciliata* is highly sensitive to allocation principles and that allocation assumptions may be a major factor influencing carbon sequestration calculations of tropical forests under climate change. Thus, carbon allocated to fine roots and leaves turns over rapidly and returns quickly to the atmosphere, whereas carbon allocated to wood is fixed for decades or even centuries (Zeng, 2008).

Another factor influencing the cycling of carbon in rainforest ecosystems is the rate of production and decomposition of forest litter. Studies over a 7-year period at the Daintree Rainforest Observatory in north Queensland (16°06′ S, 145°26′ E) have shown strong seasonality in the production of various components (leaf, wood, flowers and fruit) with peak falls coinciding with maximum daily temperatures and occurring 1–2 months before maximum monthly rainfall (Edwards et al., 2017). It was further concluded that the correlation between climate and litterfall patterns was probably consistent across the wet lowland forests of that region, making it easier to construct long-term estimates of carbon stocks and flows to support predictions of likely responses to climate change.

TROPICAL FOREST TREE HEALTH

Among the factors affecting tropical forests (and their capacity for carbon sequestration) is the condition known as patch death (or forest dieback), already affecting some of Queensland's tropical rainforests. It is most commonly associated with the soil pathogen *Phytophthora* and some other soil-borne organisms. *P. cinnamomi* was first detected in the north in the Mackay region and in areas north-west of Ingham in the early 1970s where it was linked with serious tree disease in both logged and virgin forest areas (Brown, 1999). In a survey carried out at Dalrymple Heights (on the Eungella Tableland west of Mackay), a wide range of tree species, belonging to a number of different families, was shown to be impacted. *P. cinnamomi* occurred in forest soil samples (detected by laboratory baiting assay using germinating *Lupinus angustifolius* seeds) and was also isolated from roots of affected tree species.

The symptoms of forest dieback are first visible as defoliation of the smallest twigs on the outer surface of the tree crown. Re-shooting often occurs (mainly on the larger branches and stems) and many of the shoots survive for 1 or 2 years before eventually dying. The affected trees finally become completely defoliated. Inspection of the cambium of the lower stem of such tree species confirmed that within 6 months of full defoliation, the trees were dead.

Of the factors influencing forest dieback, soil moisture is critical. Free water, essential for formation of the fungal sporangia and release of the zoospores, is clearly essential for successful infection of host tree roots. Tree damage attributed to *Phytophthora* was found to be often associated with drainage lines and with areas of water accumulation. It affects forests, heathlands and woodlands across the wetter regions of Australia. In the tropical north it has extended into high altitude forests (up to 1,500 m) such as those at Mount Bartle Frere and Mount Bellenden Ker (Worboys, 2006).

A dieback episode in a 0.5 ha permanently marked plot of unlogged forest in what is now the Girringun National Park (18°30′ S, 145°45′ E) has been described by Metcalfe and Bradford (2008). The plot was originally established to study forest dynamics relating to the important cabinet timber *Flindersia bourjotiana*. The dieback occurred during 1977–1989 and recovery was monitored up to 2005. An area of 770 m² was affected, resulting in the death of 13 trees (>10 cm dbh). The dead trees belonged to 4 of the 14 families represented in the area. The highest mortality occurred in the Elaeocarpaceae. Large trees suffered more than smaller trees, and the latter were also more likely to recover. Recovery from dieback was reflected in greater recruitment to the >10 cm dbh size class in areas that had been affected.

Another fungal disease that poses a serious threat to trees of the Wet Tropics World Heritage Area is myrtle rust (*Puccinia psidii*). Native to South America, it has in recent years been detected affecting plants across a wide area from Cape Tribulation in the north to Ingham in the south and as far west as Ravenshoe (Australian Wet Tropics World Heritage Authority, 2011–2012).

2 Wetlands, Mangroves and Impoundments

The wetland ecosystems of tropical Australia constitute a range of habitats progressing inland from coastal mangroves and seagrass beds to estuarine deltas, riparian swamps, marshes and extensive floodplains, many of them now afforded special conservation status. The comprehensive review by Finlayson (2005) has provided a valuable basis upon which management plans can be formulated to address specific environment-related issues including land use and community values as well as responses to change. Among the problems affecting management of these ecosystems is the sheer size of the areas involved and the nature of the terrain, much of it isolated and, at least at certain times of the year, virtually not accessible, even to off-road vehicles.

The first of Australia's wetlands to be listed (over 35 years ago) under the Ramsar Convention as being of international importance is the Cobourg Peninsula Marine Park, located at the northernmost tip of the Northern Territory (NT) (see Chapter 1, Figure 1.1). It covers almost the entire area of the peninsula and is managed jointly by the NT government and the Cobourg traditional owners through the Cobourg Peninsula Sanctuary and Marine Park Board. The active involvement of the local indigenous communities in management of the area is consistent with the Ramsar Convention, which sees a blend of local environmental knowledge and scientific understanding as being the key to effective land care.

Floodplain wetlands of the northern region typically experience an annual wet–dry cycle driven by the monsoonal rainfall pattern. They are created when floodwaters overflow from large streams during the wet season, but at other times of the year they generally dry out leaving a few permanent billabongs (water holes) and swamps. This strong seasonal variability of flow is greater than is experienced in any other of the world's tropical regions (Warfe et al., 2011). Northern-draining catchments have been estimated to cover approximately 17% of Australia's land mass but generate over 60% of the total surface water run-off. The biota of the floodplain wetlands is adapted to this marked hydrological seasonality so that any changes to or disruptions of this pattern are likely to have far-reaching consequences for aquatic biodiversity and a range of ecosystem processes. Many wetland areas are linked with and replenished by groundwater sources through natural springs, soaks, permanent pools and flowing vents.

UNDERGROUND WATER AND INTERMITTENT RIVERS

A large proportion of the terrain of northern Australia overlies what has been described as the lifeblood of the inland regions, the vast resource of underground water without which much of inland Australia would have remained semi-arid, with little or no agricultural development and no significant settlement. The entire system (the Great Artesian Basin) extends across 1.7×10^6 km^2 and is one of the largest of the world's underground water reservoirs. The salient features of Australia's underground water storage systems have been described by Willmott (2017a). Artesian water (free flowing) and sub-artesian water (requires pumping) from underground aquifers (mainly in sandstone beds) flows to the surface under the pressure exerted within the basin. Water enters the aquifers from intake areas and if those areas are topographically higher, a pressure head will develop creating an artesian well. In the Australian Great Artesian Basin, the aquifers occur within sandstone layers laid down in river plains of the Late Jurassic. Typical intake areas for these aquifers occur along the western margins of the Great Dividing Range and are fed by rainfall on the sandstone terrain and by river flow across it.

Artesian water originates from rainfall that percolates through the rocks to recharge the aquifers. Some of the stored water is released to the surface through natural springs, mainly around the margins and ranging in size from small soaks to large pools from which numerous flowing vents feed associated wetlands. Water from the Great Artesian Basin has historically been an important resource available via numerous free-flowing springs supporting unique ecological communities containing many endemic plant and animal species isolated from others by the largely waterless landscape. These communities represent links with the geological past before the Miocene/Pliocene drying of inland Australia. The majority of these springs are no longer active, mostly as a result of overextraction. Some survive with greatly reduced flow, while others, especially in more recent times, have suffered extensive degradation from trampling by stock and feral animals. Such open drains have over recent decades been progressively capped and the water piped to storage tanks.

Many of the rivers of the tropical north have significant flow only during the short, wet season (usually during January to March). During the long dry period there is either no or a much-reduced flow, most of it fed by groundwater seeps and springs. One such river is the Daly River in the Northern Territory, which flows north to Anson Bay in the Timor Sea (see Chapter 1, Figure 1.1). Its catchment has historically had cattle grazing as its major land use and hence a fairly low anthropogenic demand for water. In recent years, however, there have been increases in horticulture, irrigated agriculture and silviculture activities with greatly increased demand for water. Among the Northern Territory's plans for developing the region's water resources is the proposed diversion of wet season flows from the main channel of the Daly River into storages for dry-season irrigation (flood harvesting). An alternative option also under consideration is increased groundwater extraction with inevitable effects upon dry season flows in the river.

Anticipating such possible developments, there have been studies of their likely effects upon flow regimes in the river and, more widely, upon the entire riverine ecosystem and local groundwater resources. Townsend et al. (2017) have described the major benthic primary producers of the river ecosystem, comprising periphyton, microalgae, filamentous algae, the macroalgae *Chara* and *Nitella*, and the vascular macrophyte *Vallisneria*. It was predicted that the planned increases in water extraction would greatly alter the flow regime in the river as also, but perhaps in a more positive way, would any climate-change-related increased intensity and frequency of extreme rainfall events and cyclones.

NORTHERN WETLANDS

The Ord River wetland (Figure 2.1) is a prime example of the changes that development and the extensive modification of river flow have brought to what was originally a highly variable, intermittently flowing river system but has now become a perennial system with water released from the impoundment during the dry season to support irrigation and hydroelectric power demand (Warfe et al., 2011). This has

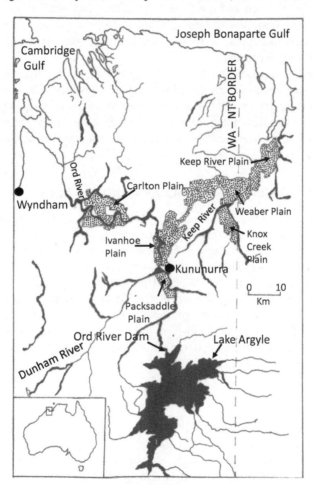

FIGURE 2.1 The rivers, impoundments and wetlands of the Ord River Irrigation Area (shaded areas) and the locations of current and predicted future areas suitable for irrigation agriculture (stippled areas). (Redrawn from the original map "Ord River Irrigation Project" published by the Government of Western Australia, Ministry for Primary Industry, 1995, and updated with permission under the Creative Commons licence of the Geographical Information Services of the West Australian Government.)

come about following construction of the diversion dam at Kununurra and the later creation of Lake Argyle. The effects on the wetland were reported as an increased sedimentation and narrowing of the river below the dam with consequent major changes to the riparian vegetation assemblages. Lowering of salinity levels in the estuary has led to reduction of banana prawn populations (Kenyon et al., 2004) and an overall loss of primary production (Burford et al., 2011).

The wetlands of the Kakadu National Park are fed by flows from the Alligator Rivers system. They were formed, according to Woodroffe et al. (1989), some 6,000 years ago after the post-glacial sea-level rise. Further substantial sea-level rises are predicted as a result of climate changes with anticipated impacts on the ecology of the wetlands and the rich and spectacular biota they support. The rising sea levels and the increasing frequency of intense tropical storm surges are a particular threat to the migrating flocks of magpie geese (*Anseranas semipalmata*), which congregate to feed on *Eleocharis* chestnuts.

The wetlands comprise a network of channels with low drainage and, even after the eradication campaigns commenced in the 1980s, are still subject to disturbance by feral animals such as water buffalo (*Bubalus bubalis*) and wild pig (*Sus scrofa*). They are prone to saline intrusion and would be particularly vulnerable to any increases in sea level. They are also subject to invasion by weeds such as mimosa and introduced pasture grasses.

BILLABONGS, FARM DAMS AND ARTIFICIAL LAKES

One of the most distinctive features of the water resource of the tropical north is that despite its abundance (accounting for approximately 70% of the country's total) its availability to support both natural and developed ecosystems is extremely variable, both geographically and seasonally. Some areas receive so little annual rainfall that river flow ceases completely for extended periods leaving isolated water holes or billabongs; others have to rely almost entirely on groundwater. Continuity of water supply can, in most cases, only be guaranteed through the construction of various forms of water storage ranging from farm dams or weirs across intermittent creeks or rivers to engineered structures large enough to create artificial lakes capable of supporting extensive irrigation schemes.

FARM DAMS

Farm dams represent a critical form of infrastructure supporting stock and domestic watering as well as limited crop irrigation. Such water use has, in the past, been largely unregulated but is increasingly coming under strict control to ensure that usage does not exceed entitlement. It is now generally accepted that some former practices such as spillway blocking and construction of excessive storage capacity can no longer be tolerated (Tingey-Holyoak et al., 2013).

The biota of billabongs and other small and often temporary impoundments are regularly subject to extreme conditions such as recurring diurnal periods of hypoxia (defined as dissolved oxygen concentrations [DO] of <2 mg l^{-1}, i.e. <26% satura-tion at 30°C). The tropical fish species *Melanotaenia utcheensis* (rainbow fish), for example, has been shown to tolerate fluctuating periods of hypoxia provided the minimum DO does not fall below ~20% (Flint et al., 2018). This makes *M. utcheensis* more tolerant of hypoxia than is generally the case for temperate fish species. At minimum diurnal DO exposure of 10%, however, both reproduction and viability of the embryos were impaired and gradual further oxygen depletion to ~7% DO has been shown to be lethal. During the day, the DO of tropical waters invariably increases, largely due to photosynthetic oxygen production by the stands of submerged plant material, the preferred spawning sites for *M. utcheensis* (and other tropical fish species). The fact that fish kills attributed to hypoxia do regularly occur in tropical waters would suggest that even the more hypoxia-tolerant species (like *M. utcheensis*) would still be vulnerable to even small increases of temperature (with reduced solubility of oxygen) or reduced flow resulting from climate change.

Weirs and River Barrages

With the exception of some volcanic crater lakes on the Atherton Tablelands (see Chapter 1, Figure 1.4), the northern region of Australia has very few large natural, permanent freshwater bodies, mainly because of the absence of significant historical glaciation (Bridgeman & Timms, 2012). Development of northern Australia, even in regions well supplied with river systems, has since its early days required the construction of weirs or river barrages to ensure continuity of supply. In the Fitzroy River basin inland of the city of Rockhampton (23°22′ S, 150°3′ E), for example, early establishment of the pastoral/wool industry and later diversification into dairy farming and various forms of cropping became possible only following the construction of weirs (and later dams) on the Dawson, Comet, Nogoa, McKenzie and Isaac Rivers (all tributaries of the Fitzroy River).

Further north, in the Bowen, Surat and Galilee Basins, inland of the city of Mackay (21°0′ S, 149°1′ E,) there are vast grazing lands, forested areas and extensive coal reserves and mining-related industrial development. The last of these, in particular, makes heavy demands on local water resources as well as leaving what is often irreversible environmental damage. Exploitation of the coal reserves of the Bowen Basin requires the construction of an extensive system of stream diversions, often with modification of river reaches and their replacement by engineered channels or pipelines to service coal mines and associated communities (White et al., 2013). The degraded nature of these catchments has drawn the attention of the State government, bringing strong recommendations for improved management of downstream impacts and greater protection of riparian vegetation and local ecosystems.

Open-cut, strip mining for coal requires onsite storage of rainwater backed up by a system of controlled releases to minimise downstream harmful effects. Post-mining mine-site rehabilitation also requires water, most critically during early revegetation stages but extending well beyond the active mining phase. More recently introduced mining techniques such as the extraction and processing of oil shale, the extraction of coal-seam gas and hydraulic fracturing to release gas reserves, all have extensive water requirements that can only be met from on-site weirs or other storages or, even more environmentally suspect, from the ever-deteriorating, and in many regions the already depleted, reserves of groundwater.

The Burdekin River catchment extending between latitudes ~19° and ~24° S, is the second largest on the east coast of Queensland. Before development of Lake Dalrymple, the delta floodplain was already a most productive area, supporting high yields of sugar cane, long-grain rice and a range of fruit crops, all under irrigation from a number of weirs constructed on local rivers. At Townsville, the largest city in far north Queensland (see Chapter1, Figure 1.4), its establishment as a centre for the pastoral industry and its later development as an administrative centre for a range of industrial, mining and community activities was, at first, completely dependent upon freshwater supplied by a series of weirs constructed along the Ross River. These were later replaced by piped water from Paluma Dam located in a rainforest area north of the city, later supplemented by the much larger Ross River Dam some 20 km south of the city and, at times of exceptional need, by water pumped from Lake Dalrymple in the Burdekin region.

Weirs constructed on streams in the Barron River region of the Atherton Tablelands (inland from Cairns) (Chapter 1, Figure 1.4) provided water for what is probably one of the earliest examples of a limited form of irrigation farming in tropical Australia. Later construction of the nearby Tinaroo Falls Dam (in the 1950s) was the first major water resource development in northern Australia and also provided a much-needed source of hydroelectric power.

ARTIFICIAL LAKES AND RESERVOIRS

Construction of large water-storage reservoirs in regions devoid of natural lakes, impacts strongly on the environment through restricting natural water flow while at the same time creating new habitats. This is well illustrated in the Mount Isa region (see Chapter 1, Figure 1.1), where exploitation of its mineral resources was made possible only through the creation of two large lakes (Lake Moondarra and Lake Julius) and a number of other smaller lakes within the catchment of the Leichhardt River (which flows north to drain into the Gulf of Carpentaria) to provide water for urban and industrial use. The environmental implications of this development have been well documented and include eutrophication (and examination of the effectiveness of various remedial measures) (Finlayson et al., 1984) and a range of other water-quality and water-supply issues (Room et al., 1981; Wrigley et al., 1991). Creation of these large lakes in such an arid region (with average annual rainfall of 200 to 400 mm) has provided suitable habitats for the establishment of substantial fish populations supplemented by an active program of fish-stocking sourced initially from local rivers and later from hatchery facilities, thus making a valuable contribution to local recreational pursuits.

The lakes of tropical Australia, ranging in size from small impoundments such as Solomon Dam ~500 ML, providing the freshwater needs of a small Aboriginal community on Palm Island, 60 km offshore from Townsville; Figure 1.2) to the ~10 × 10^6 ML maximum capacity of Lake Argyle (in the Ord River Irrigation Area of Western Australia; Figure 2.1) have all, in their different ways, presented water-supply management problems reflecting the special conditions imposed by the tropical environment. Some, like Lake Dalrymple in the Burdekin River area, because of the climatic conditions, the geomorphology of the catchment soils and the sparse vegetation cover (due in part to past inappropriate land-use practices) are subject to persistently high loadings of suspended material with consequent undesirable and potentially costly effects on use of the water for irrigation (Faithful & Griffiths, 2000).

Many of the lakes, as is typical of lakes of the warm belt, suffer extended periods of strong water-column stratification, creating conditions conducive to the development of microalgal blooms including those containing potentially toxic cyanobacterial species. Solomon Dam has been recorded as the site of one of the more serious cases of human poisoning by toxic cyanobacteria, due to *Cylindrospermopsis raciborskii* (Hawkins et al., 1985), relieved primarily by artificial mixing of the water column backed up by passage of the drawn-off water through slow sand-filtration beds followed by chlorination and treatment with activated carbon (Hawkins & Griffiths, 1993; Griffiths & Saker, 2003).

MANGROVES

The mangrove forests of northern Australia cover an area of ca. 10,918 km² (estimated from data presented in Galloway, 1982) distributed as a narrow, discontinuous coastal strip extending from Rockhampton in Central Queensland to the north-west cape in Western Australia. By far the greatest diversity of mangrove species are found along the moist tropical coastlines of northern Australia, but mangroves can be found as far south as the temperate Victorian coastline where they are represented by a single species (*Avicennia marina*). The mangroves of the tropical northern coastline are by far the most luxuriant, forming extensive tidal forests (Mitchell et al., 2007) rivalling those located in many areas of south-east Asia. They occur in tidal creeks and, in some areas they may extend inland for up to 40 km. Climate is one but not necessarily the only nor even the major factor influencing mangrove distribution.

Some examples of the long-term (over several millennia) mangrove developments in northern Australia are those established on the high-tidal flats formed in the Cambridge Gulf–Ord River region of Western Australia (Figure 2.1) and those occupying the cemented platforms that have developed on the windward sides of low wooded islands of the Great Barrier Reef. Areas experiencing rapid seaward progression of the shoreline or deposition of alluvium in estuarine channels can, however, change over a period of decades. Mangrove development is strongly affected by changes in the volume and variability of river run-off and hence the discharge of sediment or changes in the salinity of coastal estuaries, lagoons and deltas.

Mangroves are likely to be particularly affected by sea-level rises and by climatic variations. Stratigraphical records of mangrove ecosystems during the sea-level changes of the Holocene (Ellison & Stoddart, 1991) have confirmed the major influence of fluvial sedimentation of terrestrially derived material on mangrove development. Retreats of mangrove ecosystems have been related to high rates of sedimentation, which in northern Queensland sites can be as high as 270 m/100 years (Belperio, 1979). Northern Territory sites showed extensive mangrove development between 7,000 and 5,500 BP, a time of sea-level stabilisation which was, however, followed by a period of extensive sedimentation and consequent destruction of mangrove habitats. Pollen analyses of drill cores from the South Alligator River estuary and floodplain (Northern Territory) showed changes in mangrove species with marine tolerant species being replaced, over time, by more terrestrial species (Woodroffe et al., 1989).

Formation of the extensive mangrove systems of northern Australia started ~8,000 years BP when rising sea levels invaded the river valleys. Those associated with the South Alligator River estuary, for example, were strongly influenced by the relatively low catchment/floodplain ratio (9) compared with the much higher values (29) associated with another Northern Territory river system, the Daly River system. Pollen analysis of core sediment samples showed that mangrove forests were established when the sedimentary system was 8 m below the present-day plains. Mangrove was shown to persist through 7 m of vertical accretion before giving way to sedge and grassland in the floodplain clay about 1 m below the top of the core. That part of the core between 8 and 4 m represents persistent mangrove growth with rising sea level with *Rhizophora* slowly gaining ascendancy over *Brugaiera/Ceriops*. *Sonneratia* became locally important at around 3 m core depth, corresponding with sea-level stabilisation. Continued sedimentation was accompanied by successive changes from *Sonneratia*, through *Rhizophoraceous* forest and finally to stands of *Avicennia* before the mangrove was overtaken by freshwater floodplain.

It has been suggested that mangrove development might keep up with sea-level rises of up to 8–9 cm/100 years but are unlikely to persist under sea-level rises of over 12 cm/100 years, because of the low rate of sediment accumulation (Ellison & Stoddart, 1991). An assessment of the likely response of wetlands/mangroves of the Alligator River Region (NT) to the predicted climate-change-related sea-level rise has been proposed by Eliot et al. (1999). They predict extensive biophysical alteration of the coastal and wetland environment with consequent degradation or destruction of existing values as currently applied to the area such as usage by traditional Aboriginal occupants or more general nature conservation criteria. The recommended management regime recognises traditional Aboriginal ownership of much of the land (leased to the federal government as a national park) and proposes empowerment of such local groups to establish monitoring programs and to be closely involved in all management decisions.

Rehabilitation of cleared mangrove areas such as that at Port Hedland (20°1′ S, 118°3′ E) in the Pilbara region of the north of Western Australia (see Chapter 1, Figure 1.1) is now receiving considerable attention. The Port Hedland development, now under construction, will provide one of the world's largest export shipping facilities. The harbour is situated within a mangrove-fringed tidal creek system with a broad adjoining system of intertidal mud flats. Clearing and landscaping of the channels, carried out in 2012, created channel bank areas requiring stabilisation by suitable plantings of mangrove species appropriate for the region. Trials were conducted by the Pilbara Port Authority to determine the conditions necessary to achieve acceptable levels of mangrove seedling survival (Erftemeijer et al., 2018). Nursery-raised seedlings of four local mangrove species were planted in areas subjected to different levels of tidal inundation and protected by various erosion prevention measures. Survival levels (beyond the first year) of seedlings of *A. officinalis* and *Rhizophora muconata* were 10% and 40% respectively. Average survival rates for all seedlings after 36 months was 18%, which was reported as being comparable with natural survival rates in that region. Previous studies at this location using seedlings of *Avicennia marina*, reported 31% survival beyond the first year. This was deemed to be satisfactory despite being well below survival rates achieved with this species at sites adjacent to Brisbane airport (27°2′ S, 153°0′ E). It was concluded that under the semi-arid conditions of Port Hedland, growth of seedlings was slow but, subject to suitable tidal hydrology and other habitat factors, would provide significant mangrove rehabilitation over a period of a decade or so.

CARBON FLUXES IN WETLANDS

Mangroves make a significant contribution to estuarine and inshore productivity via their net primary production, represented by the difference between gross photosynthetic carbon fixation (including organic matter incorporation into tree biomass during growth) and carbon lost as leaf litter or consumed by heterotrophs. Studies of the mangroves of Hinchinbrook Island (18°2′ S, 146°1′ E) (Chapter 1, Figure 1.2), an area of approximately 50 km^2 supporting 27 mangrove species, measured litterfall and yielded values ranging from 1.42 to 3.05 g dw m^{-2}day^{-1} (Bunt, 1982). More recent estimations (Alongi, 2014) have confirmed that mangroves are among the most carbon-rich ecosystems in the world. Studies of other mangrove systems have indicated that litterfall may represent as much as 50% of total mangrove primary production indicating that considerably less becomes invested in long-lived supporting structures with consequently only a minor contribution to carbon sequestration in plant tissues.

Tropical mangroves and peat-poor swamps, unlike the peat-rich bogs of more temperate regions, do not appear to make a significant contribution to global methane emission (Matthews & Fung, 1987). Thus, studies at coastal lowlands of the Bellenden Kerr National Park (some 70 km south of Cairns in north Queensland) (Chapter 1, Figure 1.2) have shown that during the course of a year the soils are more likely to function as a significant sink for CH_4 (Kiese et al., 2003) with most of the CH_4 uptake occurring during the dry season. CH_4 concentrations in the soil were invariably below atmospheric concentrations. This contrasts with other tropical wetlands in the world, which are generally reported (e.g. Sjögerstein et al., 2014) to be important sources of CH_4 (and CO_2) reflecting their high rates of net primary productivity and litter decay. It also contrasts with global figures recently presented by Rosentreter et al. (2018) that, on a global basis, between 18 and 22% of buried carbon may be lost to the atmosphere as CH_4.

The complexity of carbon cycling in tropical mangrove ecosystems is well illustrated by the study of Robertson and Daniel (1989) describing the role of crabs in litter processing in high intertidal *Ceriops* and *Bruguiera* mangrove forests of the north Queensland coastline. It was estimated that the rate of leaf processing by crabs was over 75 times the rate of litter decay by microbial action thus facilitating the high sediment bacterial productivity of these forests.

Large areas of coastal north Queensland, the Northern Territory and northern Western Australia support extensive *Melaleuca* forests (Mitra et al., 2005) that may occupy various habitats depending upon the level and frequency of inundation ranging from continuous standing water to no or only brief inundations. Some of the forests are heavily degraded while others are undergoing some regeneration. They all have the potential to provide substantial storage of carbon estimated by Tran and Dargusch (2016) as ranging from over 300 t C ha^{-1} (in good sites) to lower values closer to 200 t C ha^{-1} in degraded or regenerating sites.

Melaleuca swamps on two islands (Mua and Badu) located in the Torres Strait, between mainland Australia and Papua New Guinea, have yielded sediment core data that provide a record of hydrological and vegetational changes dating back to 2,700 years BP (Rowe, 2015). Pollen and charcoal analyses show progression from seasonally moist–dry open herbaceous habitats to extensive stable-boundary swamps and swamp forest. The eucalypt-dominated woodlands show increasing fire influences and islander colonisation.

WETLANDS OF TROPICAL FLOODPLAINS

The tropical wetlands of northern Australia occupy the coastal fringes and the estuaries and floodplains of a series of major rivers extending from the north of Western Australia, across the Northern Territory, on to the Gulf of Carpentaria and beyond to Cape York Peninsula and south along the east coast of Queensland. They are recognised as a unique resource that, in some areas, is already under threat from various forms of development and from the effects of climate change. Concern about the possible effect of these changes has stimulated great interest in the ecology and diversity of these important ecosystems and their future sustainability. Wetlands of different regions will vary in their major components and in the extent of their exposure to and resistance to the various drivers of change.

Wetland vegetation, in its general features, has been described (Cowie et al., 2000; Finlayson, 2005) as being dominated by swamp forests capable of withstanding inundation of up to 1 m during the wet season but also able to survive long periods of little or no rain. The dominant wetland forest trees are species of *Melaleuca* above an understorey of sedges (dominated by *Eleocharis* species) and a range of other aquatic plant species. The sedgelands often merge into grasslands composed of *Oryza* and other grass species and other sedges including *Cyperus*, *Fimbristylis* and *Fuirena* species. Inundated regions support floating-leaved plants (e.g. *Netumbo*, *Nymphaea* and *Nymphoides* species) and a number of submerged and emergent species of *Urticularia* and *Limnophila*. Variations on this general picture, as Finlayson (2005) has reported, can be found across the range of northern wetlands depending on habitat differences and different degrees of disturbance. Over recent years, for example, the virtual elimination of feral buffalo (*Bubalus bubalis*) with the consequent reduction of grazing, trampling and wallowing, has allowed the reappearance of plant species formerly out-competed by invasive species.

INVASIVE SPECIES

The tropical north has, historically, suffered extensive invasions from a range of alien species introduced either deliberately (for a variety of reasons) or accidentally. The invading organisms range from bacterial, algal, fungal and insect infestations, to numerous troublesome plant and animal species. They all represent a threat to biodiversity and in many cases have resisted all attempts at eradication or control. Tropical wetlands are particularly prone to invasion because they provide habitats attractive to colonizing species, free of natural biological control agents and habitats from which they rapidly spread to other areas.

Rangeland ecosystems are also affected by invasive species, many of them deliberately introduced to improve the nutritional quality of native pastures. Some others were first planted for their aesthetic appeal or as ornamental plants in aquaria or small ponds only to spread to infest farm dams, billabongs or wetlands. Such horticultural escapes may lay dormant for decades before becoming prominent weeds with a cascade of consequent ecological effects. Some introduced plant species such as *Lantana camara* and *Cinnamomum camphora*, at first freed from natural predators, may later come to support feeding by certain native bird or insect species. Hynes and Panetta (1994) have documented some of the most serious plant and animal invaders of northern regions and their impact on agricultural production and on the environment generally. Some examples are described next.

THE SPREAD OF CANE TOADS THROUGHOUT TROPICAL WETLANDS

Cane toads (*Bufo/Rhinella marinus*) were first introduced to the cane fields of northern Australia from Hawaii in 1935 in an attempt to control the cane beetle (*Dermolepida albohirtum*). They were first released in the north of Queensland in the Cairns–Gordonvale–Innisfail area (Chapter 1, Figure 1.2), where they multiplied rapidly and spread south and to the west, reaching the Northern Territory border by 1984. By 2013 they had come to occupy more than 1.5×10^6 km^2 of northern Australia (Tingley-Holyoak et al., 2013). Among their most notable ecological effects was the poisoning of native predator species due to their defensive chemicals (bufadienolides and related toxins), which are unlike the toxins possessed by any of Australia's native animals.

Another of the consequences of the toad invasion comes from competition between the invaders and the native fauna. This can involve competition for food resources or for choice of habitat. Such competition was investigated for a location in the Leaning Tree Lagoon National Park (12°3′ S, 131°1′ E), Northern Territory, shared by the invasive cane toad and three species of native frog (*Litonia tornieri, L. nasuta* and *L. dehlii*) (Blach et al., 2014). It was found that the mature native frogs (which frequently shelter in burrows during the day) preferentially aggregated with conspecifics or with other species of native frogs. They rarely aggregated with cane toads, which also seek out shelter in burrows during the day. These results held for natural burrows and for created standardised experimental burrows.

The fact that native frogs avoided burrows containing cane toads (as well as those affected by cane toad scent) indicates that competition for suitable burrows as diurnal retreats is unlikely to be a major factor in the interaction between native frogs and invasive cane toads. Native frogs, it appears, are very selective in their choice of burrows for diurnal retreat, whereas cane toads are much less selective. Any environmental changes that alter the balance between the native frog and invasive cane toad populations and the availability of suitable refugia might, of course, have far-reaching consequences on the wetland ecosystem.

Noting the poisonous effects of cane toads upon their predators, it has been suggested that it may be possible to train local fauna (particularly endangered species) to avoid consuming cane toads (Ward-Fear et al., 2016). To be effective, such taste aversion methods would need to be conducted on a broad-scale and should affect a large proportion of the target species. Field trials conducted at a 300,000 ha property at the Mornington Wildlife Sanctuary in the central Kimberley region, Western Australia (Chapter 1, Figure 1.1), examined the extent to which the endangered northern quoll (*Dasyurus hallucatus*) could be trained to avoid consuming cane toads (Indigo et al., 2018).

When baited cane toad sausages (containing minced cane toad legs) were distributed widely across the area, they proved very attractive to wild quolls but, interestingly, 40–68% of the quolls developed an aversion to further bait consumption. It would appear, therefore, that in this particular case, at least, the impact of an invasive species may be mitigated not only by controlling the invader but by manipulating the mechanism of its impact.

MIMOSA PIGRA IN THE ANSON BAY REGION

The coastal floodplain of the Anson Bay region in the Northern Territory covers an area of 3,480 km^2 approximately 150 km south-west of Darwin (Chapter 1, Figure 1.1). Invasion by the giant sensitive plant (*Mimosa pigra*) has overgrown much of what were formerly the natural habitats for a large number of water birds, including at least 10 threatened species. Also affected are areas developed as grazing pastures as well as sites important for indigenous cultural activities. Some areas of *Mimosa* infestation are dense enough to interfere with access to fishing locations to the obvious detriment of local tourism.

A number of different control methods have been attempted with only limited success, partly because the infestations often occur on rangelands of low economic value making chemical or mechanical control methods too costly. Greater success has been achieved through the application of a more integrated approach, which includes biological control as a promising long-term management strategy supported by a range of other control options including herbicide, fire and mechanical control treatments (Paynter & Flanagan, 2004). Of the five (insect) species tested for their effectiveness in controlling growth of *Mimosa*, the stem-attacking moth *Neurostrota gunniella* showed most promise, especially when applied in combination with other control measures. Integration of different control measures allowed cost reductions (compared, for example, with chemical or mechanical methods applied on their own) without loss of effectiveness.

INVASIVE TREE, ZIZIPHUS

Studies carried out at the floodplains of the Forrest River in the Kimberley region of northern Western Australia (15°0′ S, 127°5′ E) (Chapter 1, Figure 1.1) provide a good illustration of the complexity of the interactions between different components of an ecosystem already under various forms of environmental stress. The floodplains provide a critical refuge habitat for a native rodent species *Rattus tunneyi* (pale field rat) in a region much affected by the trampling of feral horses (*Equus caballus*) and by the invasive tree species *Ziziphus mauritiana* (chinee apple).

It was noted (Ward-Fear et al., 2017) that rat burrows were twice as common under chinee apple trees as they were under most other tree species. The burrows under chinee apple trees were, moreover, home to more than seven times as many native rats as occurred under other tree species, and, unlike the smaller burrows under the other trees (which contained only male rats), the chinee apple burrows contained both male and female rats. Clearly, the invasive tree species, perhaps because of its thorny foliage and hence effective exclusion of feral horses from its vicinity, plays a critical role determining the persistence of the pale rat population in this degraded ecosystem. The recommendation for management of this ecosystem was that *Ziziphus*, despite its alien origins, should be retained, at least until the feral horses could be removed from the area.

RUBBER VINE (*CRYPTOSTEGIA GRANDIFLORA*)

A landholder survey conducted some 25 years ago estimated that rubber vine (an invasive species originally from Madagascar), was well established across some 700,000 hectares of Queensland (Hynes & Panetta, 1994). Since that time, it has continued to spread across the north and now grows in many areas of the north of Western Australia (Palmer & Vogler, 2012). It is recognised as one of the Weeds of National Significance, although during World War II, after the fall of Singapore, it was planted as a possible source of latex for rubber production. It has a deleterious effect on the grazing industry by reducing pasture yield and interfering with livestock management. It also produces a milky latex containing cardiac glycosides that are toxic to livestock. Feeding tests have shown that the leaves are toxic to cattle, sheep and goats with horses being particularly susceptible (Everist, 1981). A range of control measures have been applied with varying degrees of success but all deemed to be uneconomic (Vitelli, 1995).

A search in rubber vine's native Western Madagascar location for possible biocontrol agents revealed two possible agents (Palmer & Vogler, 2012). One was the leaf-feeding moth *Euclasta whalleyi* and the other was an autoecious rust fungus *Maravalia cryptostegia*. The moth, when released to rubber vine under quarantine conditions, survived but took many years to cause extensive damage to the rubber vine leaves. The rust fungus, on the other hand, established very rapidly throughout the rubber vine's range and proved to be very effective in providing a high degree of control. When ranked (on the basis of a cost-benefit analysis) against other Australian biocontrol projects, the biosecurity offered by *Maravalia* was placed third behind two other control projects (prickly pear and skeleton weed). Another beneficial effect of the rust biocontrol was that it induced leaf fall, allowing grass to grow beneath the rubber vine, providing fuel for prescribed burning and thus increasing the efficacy of burning as a control method. Research conducted at the Tropical Weed Research Centre, Charters Towers (Chapter 1, Figure 1.4), confirmed that biocontrol of rubber vine is most effective when applied as one of the elements of an integrated program of control involving chemical, mechanical, fire and biological methods (Vitelli, 1992).

Salvinia molesta, a Floating Aquatic Fern

Salvinia molesta is one of a number of waterweeds (the others include the rooted species *Hydrilla verticillata* and *Potomageton crispus*) that started to become established in Lake Moondarra, Mount Isa (see Chapter 1, Figure 1.1), in the early 1970s (Farrell, 1978). It was first described in Lake Kariba, Zimbabwe (Mitchell & Turr, 1975), where, because of its rapid rate of colonisation of suitable areas of water, it was deemed to be comparable with another notorious floating weed *Eichornia crassipes*. In Lake Moondarra it was estimated (Farrell et al., 1979) to cover an area of over 330 ha confined to the southern sections of the lake by a boom constructed across the lake at the pump station. Its estimated biomass at the time was in excess of 53,000 t fresh weight with a density cover of 167 t ha^{-1}.

Attempts to control the growth of *Salvinia* in Lake Moondarra by spraying with herbicides proved costly and largely ineffective. Biological control using the beetle *Cyrtobagous singularis* was, however, much more successful with extensive damage recorded to the leaves and buds of *Salvinia*. Within a year of the initial release of the beetle populations (and with further periodic releases) the damage to *Salvinia* was extensive leaving very little of the fern mass unaffected. Moreover (and unexpectedly) the *Salvinia–Cyrtobagous* association seemed to have stabilised at low beetle population densities perhaps, according to Room et al. (1984), because of the fortuitous choice of a race of *C. singularis* particularly adapted to *S. molesta*.

Rubberbush (Calotropis procera)

Rubberbush is a coarse shrub, native to tropical Asia and Africa (Everist 1974) which has become naturalised in the Kimberley district of Western Australia (Chapter 1, Figure 1.1), northern areas of the Northern Territory and a number of locations in northern Queensland (Chippendale & Murray 1963; Carruthers et al., 1984). Feeding trials on sheep fed with a basal ration of cracked rice and chaffed legume hay supplemented by chaffed fresh *Calotropis procera* leaves and flowers (30 days at 5 g·kg^{-1} body weight, followed by a further 30 days at 10 g·kg^{-1} body weight) showed no clinical symptoms nor, at post-mortem, any significant gross pathological or microscopic abnormalities (Radunz et al., 1984). Similar trials with weaner Brahman cross-bred cattle yielded similar results confirming the observations of graziers in northern Australia that *C. procera* could be safely used as a protein supplement for cattle in the dry season when native pastures are low in crude protein. *C. procera* has also been considered as a possible source of biocrude (i.e. hydrocarbons plus some other non-polar compounds), although the yields so far obtained are generally well below those from other sources such as seed oils (Carruthers et al., 1984).

Introduced Vermin

Growing wheat and other crops, especially at locations such as Kununurra and the Ord River Irrigation Area (Figure 2.1), can only be successful if strict measures are taken to control vermin such as mouse (*Mus domesticus*) plagues and rats such as *Rattus sorditus* and *R. villossissimus*, which have been reported to cause substantial losses to dry-season crops such as melons, sunflower and maize. The standard pest control for such infestations is treatment with the poison sodium fluoroacetate (1080) but only after prior testing of the sensitivity of native non-target fauna.

Such testing was carried out on nine species of native animals from the north-western Australian region using the increasing dose procedure to determine the approximate 1080 lethal dose for each species (Martin & Twigg, 2002). It was found that granivorous birds (e.g. ducks and corellas) from the region were generally more sensitive to 1080 than their counterparts from southern Australia. Birds of prey from the area were, however, more tolerant of 1080 and would, therefore, face little risk of secondary poisoning from 1080 used in pest control programs aimed at rodents or rabbits. It was further confirmed that risk of primary poisoning of raptors (such as brown falcons or barn owls) from meatballs containing 6 mg or less of 1080 per bait was very low.

Concerns such as these, especially the potential risk associated with the use of baited grain, led to trials to test an alternative poison zinc sulphide (ZP) whose toxicity is due to the release of phosphine gas (PH_3) or phosphorus oxyacids (Twigg et al., 2001). Trials carried out at the Frank Wise Institute Research Station at Kununurra applied ZP in the form of baited whole grain to plots of irrigated crops at times when rats cause most damage, i.e. at the end of the dry season and the end of the wet season. It was confirmed that, depending on the method and timing of bait application, loss of ZP was minimal with enough of the poison remaining in the wheat grain after 8–14 days to be lethal to rats. Other studies (e.g. Staples et al., 2003) have confirmed the relative safety of ZP for secondary non-target species and its low persistence in crops, soil-water or the atmosphere, making it suitable for inclusion in any integrated pest-management strategy for rodents.

3 Tropical Savannas

Much of the inland tropical north of Australia is covered by sparsely wooded savanna (see Chapter 1, Figure 1.3), with annual monsoonal rainfall averaging from 200 mm in the southern region of central Australia to 800 mm and above closer to the northern coastal region (see Chapter 1, Figure 1.1). The region is generally characterised by low productivity levels with inherently large and unpredictable intra- and inter-annual variation of rainfall (D'Odorico & Bhattachan, 2012). The primary drivers of the vegetation patterns of the different regions are water availability and soil nutrients, with fire and grazing (by livestock and native herbivores) as secondary drivers (Kutt & Woinarski, 2007).

Large areas of northern Australia's savanna country owe their origin to the destruction of mesotrophic forest. They are typically dominated by grasses and are often strongly affected by seasonal changes of rainfall. C-4 species are well represented, reflecting their tolerance of dry, open habitats with high irradiation. Overgrazing or unsuitable land management has, in certain areas, increased the pressure on these already vulnerable ecosystems leading to irreversible degradation. Evergreen eucalypt species tend to dominate, often co-occurring with several deciduous and semi-deciduous tree species (Bowman & Prior, 2005). In the Darwin region, for example, two eucalypt species (*Eucalyptus tetrodonta* and *Eucalyptus miniata*) occur as co-dominants (Brooker & Kleinig, 2004). More poorly drained sites often support islands of monsoonal forest and *Melaleuca* swamps (Franklin et al., 2007).

A comparison of the flora from different localities within the savanna landscape was carried out by collecting seedlings from each of three habitat associations and growing them in a common plot (Orchard et al., 2010). It was found that species from within monsoonal forest patches had higher specific leaf area, lower photosynthetic capacity and lower photosynthetic light compensation points compared with open savanna species or those from the *Melaleuca*-dominated swamps. They also required lower irradiance levels to achieve 50% of light-saturated photosynthesis. All these traits, it was argued, probably contribute to the greater shade tolerance of species beneath the dense monsoon forest canopies. Plants from the *Melaleuca* swamps had high stomatal conductance and hence higher CO_2 uptake during photosynthesis and, inevitably, a high transpiration rate. Such a low water-use efficiency could be tolerated in the swamp environment where the high transpiration rate would bring the incidental benefit of an increased nutrient transport.

VEGETATION OF THE TROPICAL SAVANNAS

The northern savanna country (occupying up to 200,000 km^2) extends from the north-west of Western Australia across a large area of the Northern Territory and as far east as the Gulf of Carpentaria (see Chapter 1, Figure 1.3). It represents the

world's most extensive and intact eucalypt open forest comprising an open overstorey canopy of less than 50% cover, an under-storey of semi-deciduous small trees
and shrubs plus a seasonally continuous cover of annual and perennial C-4 grasses
(such as *Sorghum* spp.) dominating the under-storey (Mott et al., 1985). The more
northern regions can receive annual rainfalls in excess of 1,200 mm. The more southern regions are seasonally dry and are, therefore, suitable only for plants capable of
surviving long periods of arid conditions.

Because of the highly variable climate conditions of large areas of the savanna
country, the pattern of leaf production and leaf fall is also strongly seasonal, and
many species are deciduous or semi-deciduous during the dry season. There are also
large seasonal variations in photosynthesis per unit area. Mucha (1979) showed that
stem growth of mature trees of *E. tetrodonta* growing at sites near Darwin was almost
entirely confined to the wet season. More detailed studies from the same region
(Prior et al., 2004) compared growth rates of tree species from four habitats – open
forest, mixed eucalypt woodland, *Melaleuca* swamps and dry monsoon rainforest.
For most of the species, increases in diameter at breast height (dbh) were confined to
the wet season (November to May). The average annual increases in dbh were highest in trees in the dry monsoonal rainforest (0.87 cm), then in the Melaleuca swamp
(0.65 cm), and lowest in woodland (0.20 cm) and open forest (0.16 cm) sites. Non-
Myrtaceous tree species had higher annual growth rates than Myrtaceous species
(0.53 cm versus 0.25 cm), although the latter accounted for 60–80% of the standing
biomass, probably because of their greater tolerance of fire.

Some species have been shown to possess the water-conserving crassulacean
acid metabolism (CAM) type of photosynthesis (as described earlier in Chapter 1).
Reporting on the CAM and C-4 capacity of some plants in the northern Australian
savanna country, Winter and Holtum (2015) found that in some plants of the more
arid savannas, the CAM mechanism is more cryptic as is the case of *Jatropha curcas* (Euphorbiaceae), a drought-tolerant shrub or small tree native to the American
tropics but now naturalised (or invasive) to many other tropical regions including
northern Australia. The possible use of *Jatropha* as a bio-energy feedstock has generated speculation as to whether there might not be a CAM contribution to its ability
to withstand such severe water limitations. It was observed that *Jatropha* displayed
small increases in nocturnal acid content (especially in the stems), which was interpreted as a low-level CAM characteristic. It was concluded that although C-3 photosynthesis was the principal pathway of carbon fixation in *Jatropha*, the conservation
of carbon (rather than water) in the stems via the CAM-type mechanism may contribute to the plant's well-established high degree of drought tolerance.

SAVANNA SOIL CARBON

The pool of soil carbon in savanna ecosystems is a major component of the global
carbon cycle. It has the potential to be a contributor of CO_2 to the atmosphere (from
respiration of its biota) or it may act as a sink for the products of carbon sequestration (because of its overall carbon deficit). The latter process involves the transfer of
atmospheric CO_2 into soil carbon either by humification of photosynthetic biomass
or through the formation of secondary carbonates. For most ecosystems, soil organic

carbon (SOC) represents the most stable carbon (C) pool. Changes to this pool are generally slow and may occur over decadal time scales driven by changes in vegetation cover and land use. In savanna ecosystems, long-term changes in SOC are largely due to fire regimes and grazing pressures.

Rates of C sequestration vary among different ecosystems and values between 100 and 1,000 kg C ha^{-1} yr^{-1} have been quoted for soil organic carbon and 5–15 kg C ha^{-1} yr^{-1} for soil inorganic carbon (Lal & Follett, 2009). Keeping the pool of soil carbon above a critical level is essential for sustainable agronomic production and for the maintenance of healthy ecosystems. It is particularly important in regions with fragile soils and harsh climates. Generally, savannas have low SOC compared with tropical forests or temperate grasslands, but data from savanna sites in the Darwin region have been shown to have higher SOC than are typically found in savannas (Chen et al., 2005).

Any estimation of the significance of the soil component in the carbon balance of the savanna ecosystem requires data from a range of interrelated processes. Such data have been reported for four eucalypt open forest study sites representative of the vegetation of coastal mesic savanna of the Northern Territory (Chen et al., 2003). The total C stock was estimated as 204 ± 53 t C ha^{-1}, with ~84% below ground and ~16% above ground. Soil organic carbon content (0–1 m) was 151 ± 33 t C ha^{-1}, accounting for ~74% of the total C in the ecosystem. Vegetation biomass was estimated at 53 ± 20 t C ha^{-1}, 39% in the roots and 61% in the above-ground component (trees, shrubs and grasses). Annual gross primary production was 20.8 t C ha^{-1} (27% above ground and 73% below ground). Annual net primary production was 11 t C ha^{-1} (with similar distribution between above- and below-ground components). Annual soil carbon efflux occurred at 14.3 t C ha^{-1}. Interestingly, in this region of alternating wet–dry seasons, approximately three-quarters of the carbon flux occurred during the (5–6 month) wet season, when the savanna ecosystem acted as a C sink. During the dry season, it was a C source, especially under the influence of the frequent fires when the C losses would be considerable.

Studies of $\delta^{13}C$ (see Chapter 1, section "CAM Photosynthesis in Tropical Rainforest Epiphytes") of the organic carbon pool in surface forest and grassland soil samples from different areas of northern Australia (Bird & Pousal, 1997) showed small variations in $\delta^{13}C$ values on both local and regional scales. The major source of this carbon is photosynthetic CO_2 fixation and most of it enters the pool of SOC in the surface soils. Forest surface SOC had average $\delta^{13}C$ values of $-28.4 \pm 0.7‰$; tropical grassland soils (in which C-4 plants are often dominant) gave variable $\delta^{13}C$ values, some approaching the high values (up to $-12‰$) normally associated with SOC derived from C-4 plant sources. Integrated regional $\delta^{13}C$ values for SOC can be used as a proxy for terrestrial carbon storage, with river sediments generally reflecting the values obtained with regional soils (but with a bias towards the C-3 derived values). In mixed C-3/C-4 biomes, C-3–derived carbon was noted to be preferentially incorporated into the course-sized fraction, while C-4–derived carbon was preferentially incorporated into the finer sediments.

Photosynthetically fixed carbon can also enter the soil as leaf or root litter or as root exudate, there to be subject to attack by shredder organisms and further decomposed by soil organisms and microorganisms, releasing most of the carbon as

CO_2 (Kramer & Gleixner, 2007). It is generally assumed that only the more stable carbon products such as lignin (and to a lesser extent, cellulose) will be preserved to form pools of organic matter with turnover times varying from a few years to tens or even hundreds of years. However, measurement of isotope signatures of specific compounds (Gleixner, 2013) has suggested that most soil carbon is not in the form of plant-derived residues like lignin but is derived from microbial synthesis. The carbon storage capacity of any soil will be governed partly by soil mineralogy and partly by the conditions as they affect microorganismic and microbial action.

Comerford et al. (2015) assessed the economics of devoting various forms of forestry plantations in Queensland to carbon sequestration. They compared mono-culture hardwood plantings, mixed species environmental plantings and managed forest regrowth and noted that, since most of the available area was in the Brigalow (*Acacia harpophylla*) belt, it was unlikely that there would be a significant over-all contribution to carbon sequestration. To make such areas economically viable for carbon forestry, they estimate, would require a carbon price of A$30 per tonne of CO_2 (compared with the price of approximately A$23 applied when the pricing mechanism was introduced in 2012). The Wet Tropics Management Authority has, in collaboration with a number of research establishments in the region, embarked on a program to accelerate the recovery of degraded land with a view to improving species richness and, eventually, to make a more significant contribution to carbon sequestration and storage.

Biochar, a long-lived form of carbon derived from biomass, has the potential to act as a carbon sink that can persist in the soil for thousands to millions of years (Graetz & Skjemstad, 2003). It can exist in various forms ranging in complexity from graphite-like carbon to high molecular weight aromatic compounds such as can be produced from woody wastes by pyrolysis to form charcoal (Antal & Grønli, 2003). Aboriginal cultures from other tropical regions of the world have traditionally boosted productivity of their highly weathered soils through the incorporation of biochar. Major et al. (2010) have reported improved crop yields following the application of biochar (black carbon) to their Colombian savanna oxisol. Such soil improvements have yet to be extensively trialled for Australian tropical savanna soils.

FIRE MANAGEMENT

The occurrence and severity of wildfire is a crucial component of the pre-history of the Australian environment, and analysis of charcoal from sedimentary records has provided an insight into the past occurrence of fire. Palaeo-fire records show a strong relationship between biomass burning and climate (temperature and mois-ture). In wet climates, productivity of the vegetation may be high, but the plant mate-rial and litter may be too wet to combust. In dry and colder climates, productivity is reduced and fuel loads become limiting. Australia's palaeo-fire records show con-sistent responses to climate change, with cold intervals characterised by less fire and warm intervals by more fire (Mooney et al., 2012). Records for Australia generally show that fire responses to climate change are fairly rapid. It is predicted, therefore, that the expected increases in temperature during the 21st century may well lead to a rapid increase in biomass burning.

Such responses will, of course, be strongly influenced by other factors such as those related to human activities and changing land-use practices. In tropical Australia, increases in biomass burning after about 1800, despite the reported fire suppression deemed to be characteristic of the European colonisation of Australia (Pyne, 1991), have been interpreted to be due more to post-industrial changes in climate and atmospheric composition rather than to human activities (Mooney et al., 2011). This contrasts with the 20th century decrease in biomass burning registered in many other parts of the world (Wang et al., 2010), generally interpreted to be a consequence of landscape fragmentation due to agricultural expansion. This implies that any decrease in landscape fragmentation (such as might come from reforestation for carbon sequestration) might further exacerbate climate-related increases in biomass burning (Williams et al., 2004).

Many of the tree species of mesic savannas have a strong re-sprouting capacity, which has an important bearing on their fire survival. Re-sprouting after fire may occur from dormant buds on the trunk or from a mass of dormant buds that form swellings, lignotubers, at the base of the stem. Re-sprouting exposes such trees to what has been described as the fire trap. Attempts have been made to define re-sprout curves to arrive at a stable persistence equilibrium that represents the size of individual plants upon which a population will converge over successive inter-fire time steps under a given fire regime. Fire experiments carried out at Kapalga (12°34'S; 132°19'E) within the Kakadu National Park (Northern Territory) (Williams et al., 1999) and on Tiwi Island (Richards et al., 2012) examined the dynamics of re-sprouting, while Freeman et al. (2017) devised a re-sprout curve that suggested an unstable equilibrium representing the size above which individual plants can escape the fire trap. Tests with different species from the locality (*Corymbia nesophila*, *E. miniata*, *E. tetrodonta*, *Erythrophleum chlorostachys*, *Planchonia careya*, and *Terminalia fernandiana*) allowed the construction of re-sprout curves to define the fire trap for different species under different fire regimes.

Large areas of Australia's tropical savanna country have been identified as being under covenant and declared to be nature refuges. One such area is the Bimblebox Nature Reserve (23°46'S; 146°35'E) west of Emerald in central Queensland, an area of 80 km² of private protected semi-arid country gazetted in 2003 and located in what has been described as a national biodiversity hotspot (Adams & Moon, 2013). The region has a mean annual rainfall of 517 mm and the dominant tree species is *Eucalyptus melanophloia* (10–30% canopy cover). The area is subject to occasional fire outbreaks, although there is usually not enough fuel (mostly tree litter) to burn every year. Studies by Fensham et al. (2017) covering a period of 8.5 years compared growth in plots that were experimentally burned once a year with those burned twice a year or not subjected to any experimental burning. For all three treatments, the rate of top-kill of *E. melanophloia* was relatively low. The majority of juvenile trees (<1 cm diameter at dbh) either died during the dry period or persisted as juveniles throughout the monitoring period. Girth growth was enhanced during wet years, especially for larger trees (>10 cm dbh), but all trees, regardless of size or the extent of woody competition, suffered from drought-induced mortality. It was concluded that rainfall (in particular, drought-induced mortality) was a much stronger influence than fire on the tree demographics of *E. melanophloia* in this semi-arid savanna.

It is unlikely, therefore that fire can be used to manipulate populations of *E. melano-phloia* for pastoral management. It is equally unlikely that such semi-arid savanna locations would be effective in carbon sequestration.

SAVANNA BURNING AND CO₂ EMISSIONS

Fire is well known to have profoundly influenced the historical evolution and present-day biota of the savanna lands (Kershaw et al., 2002; Williams & Gill, 1995) and prescribed burning practices have always formed a part of Aboriginal land management (Crowley & Garnett, 2000). Introduction of European-style agricultural practices and adoption of regimes dictated by pastoral land-use have brought marked changes to savanna vegetation and faunal assemblages (Bowman, 2001; Crowley & Garnett, 2000). Studies based on drier savanna regions (Kutt & Woinarski, 2007) have shown that conversion to pastoral farming regimes brought a massive influx of exotic herbivores within and beyond fenced areas, a loss of ground cover, a diversion of primary productivity and a decline in the small mammal population.

Grazing pressure impacts recorded at an arid study area located some 300 km south of Charters Towers (20°05'S; 146°16'E) (see Chapter 1, Figure 1.4) were reported to have contributed to the nature of the dominant vegetation, namely an open *Eucalyptus similis* woodland, suitable for large cattle properties with long-term paddocks (Smith, 1994). Certain areas, however, remained ungrazed due either to the presence of the shrub *Gastrolobium grandiflorum,* a native plant known to be toxic to cattle, or to dominance of the ground cover by the perennial highly flammable hummock grass spinifex (*Trodia pungens*). Both fire and grazing had a significant effect on the vegetation; they both reduced ground cover and increased its dominance by tussock grasses (*Aristida* sp.) and other (non-grass) herbs (*Phyllanthus* sp.) and shrubs (*Acacia* sp.) (Kutt & Woinarski, 2007).

Fire regimes strongly impact the exchange of greenhouse gases (GHG) between the biosphere and the atmosphere, and greatly affect the capacity of terrestrial ecosystems to store carbon. Of particular relevance are the scale and intensity of the fire and the length of the interval between fires. The widespread savanna country of the tropical north of Australia, covering an area of approximately $2 \times 10^6 \, km^2$, has been estimated to account for ~30% of Australia's terrestrial carbon stocks (Williams et al., 2004). It is, however, regularly subject to extensive fires, especially during the usually prolonged dry season (Fensham, 2012). It is these fires (and particularly their timing) that are the major drivers of fuel consumption and the resultant loss of carbon to the atmosphere (Beringer et al., 2007). Annual burning of northern Australian savanna has been estimated to release 1.6–2.9 t C ha⁻¹ year⁻¹ (Cook et al., 2005) with a net ecosystem productivity of 3.5–5.0 t C ha⁻¹ in the intervals between fires (Beringer et al., 2007). Based on these figures the savanna was estimated to represent a net sink of approximately 2 t C ha⁻¹ year⁻¹. Across all land types, however, carbon gains in one site are often offset by carbon loss in other sites resulting over the whole of the Australian savanna biome and over decadal time scales, a net biome productivity close to zero, although strongly variable depending upon climatic conditions (Barrett, 2011).

Simulation models have confirmed that, over a number of years, mesic savannas may be carbon sinks or carbon sources, depending on fire frequency (Liedloff & Cook, 2011). For such savannas in northern Australia, there is an optimal fire regime of one low-intensity fire every five years for maximum soil carbon storage (Richards et al., 2011). Generally, a reduction in fire intensity may result in higher carbon storage irrespective of fire frequency.

Aerial photographs of a savanna-forest mosaic region of the Australian monsoonal tropics taken over the period 1941 to 1994 have been compared with more recent vegetation studies to confirm the wide expansion, over time, of closed forest with accompanying contraction of grassland patches (Banfai & Bowman, 2005). The changes have been attributed largely to the cessation of Aboriginal fire management, although climatic changes may also have contributed. Savanna burning as a contribution to greenhouse gas abatement has been seen by state and territory authorities as an opportunity for some of the more remote Aboriginal communities of northern Australia to engage with the mainstream economy while, at the same time, fulfilling their traditional obligation to land stewardship.

The Tiwi Island carbon study (Richards et al., 2012) had as its main aim the identification of the biophysical and economic potential of aboriginal style fire management in the island's ecology. Satellite imagery studies over the period 2001 to 2010 have shown that, on average, ~35% (187,000 ha) of the savanna woodlands and open forests of the island is burned every year, with ~72% of the burning occurring late in the dry season (August to November). Applying aboriginal fire-stick land management (approved under federal government carbon farming initiatives) was estimated to reduce CO_2 emissions by approximately 70% compared with unmanaged savanna, with changes in fire frequency and intensity as the major contributing factors. It was further suggested that fire management might also have the potential to increase carbon sequestration in soil and vegetation.

DESERT ECOSYSTEMS

The more arid regions of the savanna country may be defined as having soils that for most of the year are under a water deficit. That is, the rainfall contribution to soil moisture is below the demands from evapotranspiration of the vegetation or from direct evaporation from soil or surface water.

Desert ecosystems, such as those occupying vast tracts of northern Australia, might be expected to rank at the forefront of vulnerability to global climate change because their biota, already existing close to their biological limits, are also exposed to the more extreme climatic changes. There is, however, some limited mechanistic evidence (Tielbörger & Salguero-Gómez, 2014) suggesting a surprising resilience of desert vegetation, particularly in relation to changes in precipitation and CO_2 concentration, probably due to specific adaptations that have evolved as a response to the stressful and highly variable climatic conditions.

It has been estimated that the predicted 3°C warming over the coming century may lead to a doubling of the frequency of climatic extremes in drylands (IPCC, 2007). It is likely, however, that the predicted (and experimentally simulated) scenarios of climate change may lie within the boundaries of those climate changes (e.g. rainfall fluctuations) already experienced by desert plant communities (see the data of D'Odorico & Bhattacham, 2012). Thus, even very drastic manipulations (such as the 75% reduction of ambient rainfall applied in the calculations by Evans et al., 2011, over a 7-year period) appear not to impose any detectable impact upon the dominant arid plant species composition. Instead, desert plants often exhibit a range of adaptations to assist their survival such as greater succulence, the development of water-saving photosynthetic pathways, leaf and stem modifications, or avoidance strategies such as annual habits, ephemeral aboveground plant parts and fine root production.

4 Tropical Crops

The coastal Queensland region of the tropical north (south of Cape York) has been a successful crop-producing area for over 100 years, with sugar cane as the major crop. Other valuable crops include tobacco (for ~50 years from 1940) and a wide range of tropical horticultural crops such as banana, mango and many soft fruits. Water storage impoundments such as those at Mareeba and in the lower Burdekin were specifically developed to support these crops, highlighting the limitations of the natural environment of the north for reliable year-round crop production. More recent years have seen a steady diversification into horticulture, aquaculture, dairy production and a range of other crops, most of them dependent upon irrigation.

Other parts of the tropical north have, until fairly recently, been far less successful for large-scale crop production, with cotton production in the Ord River Irrigation Area (see Chapter 2, Figure 2.1) during the 1960s and 1970s being the only example of successful broadacre agriculture, although that was not without a number of setbacks in the early years. Areas inland of the Great Dividing Range and across to the Northern Territory and the north of Western Australia, previously largely underdeveloped for cropping or intensive animal production, did not see any marked increase in agricultural production until the 1990s. Over the decade 1990 to 2000, however, the value of crop and livestock production more than doubled in the Western Australia–Kimberley region (see Chapter 1, Figure 1.1) (not counting long-term crops such as sandalwood), increased by almost 5 times in the Northern Territory and by 1.4 times in northern Queensland (Yeates et al., 2013). The most profitable agricultural activity across that region is still the pastoral industry but with a gradual increasing contribution from a growing range of agricultural and horticultural crops.

Climate change effects, as they apply to the tropical north, have the potential to impact on crop productivity and may require special measures to maintain profitability. Any increase in the incidence of cyclones, for example, would call for the adoption of more costly tilling methods including contouring to reduce run-off. Thus, on the slopes of the East Palmerston banana plantations and along the banks of the North Johnstone River (west of Innisfail; see Chapter 1, Figure 1.2) certain measures (supported by Natural Research Management grants) are already in place to minimise soil loss. These include laser levelling, contouring and the establishment of sediment traps and grass inter-rows. An Australian "Ag Force" report has predicted reduced profitability and additional insurance costs that would inevitably follow any increase in the incidence of extreme weather events.

There may, of course, be opportunity for crop improvement either through breeding or molecular genetics assisted by shoot imaging systems to detect dynamic phenotypic responses (Nielson et al., 2015). Another approach would be to expand cultivation of crops known from other parts of the world for their greater resilience to heat and drought stress. One such crop would be cassava (*Manihot esculenta*), which has already been the subject of a study by Meat and Livestock Australia, a body which identified certain areas of the Northern Territory as being suitable for the establishment of an integrated cassava-based stockfeed industry.

THE AUSTRALIAN SUGAR CANE INDUSTRY

Sugar cane production is a major component of Australia's rural output and makes a significant contribution to the country's export earnings. Cane is grown along a more than 2,000 km stretch of the east coast of Queensland (and extending into northern New South Wales), well over a half if it in the tropical north. Many of the cane farms are rain-fed but more recently there has been considerable expansion into irrigated areas such as the Burdekin River Irrigation Area (in Queensland) and the Ord River Irrigation Area (in Western Australia), although the latter is subject to various commercial consideration with ethanol production among the options to support possible future expansion.

Most of the cane farms are privately owned and operated. They range in size from ~30 ha to more than 90 ha (1 ha yielding, on average, ca. 84 t of cane from which about 11 t of raw sugar is produced; Australian Industry Commission, 1992). Expansion of the area under irrigated cane often comes at the expense of natural wetland areas, but in the Burdekin area at least, some cane farmers, well aware of the protective and buffering properties of wetland ecosystems, have set aside certain areas of their land for the establishment of artificial wetlands. In one case, support funding from the Australian Government's Reef Rescue program and additional support from the local water management body (North Queensland Dry Tropics) have allowed construction of a clay-lined water-cleansing lagoon (to minimise contamination of the water table) and a 3 ha wetland area (Cogo, 2010). The lagoon is a modification of the recycling pits routinely used by cane farmers to catch run-off from the paddocks for redistribution back on to the crop.

GREEN CANE TRASH BLANKETING (GCTB)

The cane crop normally grows for 12 to 16 months before harvesting (usually between June and December, when the sugar content is high). After harvesting, the stubble produces new shoots and develops a follow-up ratoon crop (usually two but sometimes as many as four ratoon crops) before ploughing and replanting of a new crop (after a 1-year fallow period). Burning of cane before harvesting was formerly a long-standing practice but is increasingly being replaced by green cane harvesting.

With this harvesting method (GCTB), the cane leaves, leaf bases and other non-harvestable residual plant material, instead of being burned, is deposited on the soil surface as an undisturbed layer and subsequently subjected to shallow cultivation. It has been estimated that well over 70% of the Australian crop is now harvested by the GCTB method (Kingston & Norris, 2001).

Trash burning leads to loss of organic matter and a considerable loss of nutrients. Field trials carried out at various Australian cane farms (Robertson, 2003) have compared the cycling of carbon and nitrogen in sugar cane soils supporting crops managed either by burning or by GCTB. It was shown that GCTB returns

7–12 t ha^{-1} of trash dry matter to the soil with almost all of it decomposing over the subsequent 1-year period. The decomposition rate, it was noted, was influenced by rainfall and temperature. Long-term studies showed that soil organic C and total N content was higher under GCTB than under burning, most of the effect being observable in the top 5 cm. The improved C availability under GCTB was observed to support increased microbial activity with most of the trash C metabolised and lost from the system as CO_2. Interestingly, there was no significant increased net mineralisation of soil N by microbial action and it was calculated that, with standard fertiliser applications, the entire trash blanket could be decomposed without affecting the supply of N to the crop. Calculations of possible long-term effects suggested that GCTB could result in a 3% to 23% increase in soil C and N depending on soil and climatic factors and that it could take 10–35 years for GCTB soils to reach a new equilibrium after conversion from burning to a GCTB trash management system.

AUSTRALIAN SUGAR CANE AND C-4 PHOTOSYNTHESIS

Sugar cane was well established as an important tropical crop yielding a valuable and saleable product (sucrose) long before the detailed mechanism by which 12 molecules of CO_2 are photosynthetically fixed to form one molecule of sucrose $(C_{12}H_{22}O_{11})$ had been worked out. That only occurred in the mid-1960s when first H. P. Kortschak and his co-workers (1965), working in Hawaii, and then M. D. Hatch and C. R. Slack (1967), working in Australia, discovered that sugar cane had a mechanism for the photosynthetic fixation of CO_2 different from that which at the time was known to occur in all plants. In sugar cane, where photosynthesis is very rapid and very efficient, supplying the radioactive form of carbon dioxide ($^{14}CO_2$) to leaves in the light yielded, as the first radioactively labelled product, two organic acids, malic acid and aspartic acid, both 4C acids in which one of the carbon atoms in each acid molecule was the radioactive form (i.e. the original ^{14}C from the supplied carbon dioxide). Expansion of this work confirmed that many other grass species of tropical origin (including two other important agricultural crop species, maize and sorghum) showed a similar pattern of early labelling of 4C acids and were hence referred to as C-4 plants to distinguish them C-3 plants which have the 3C phosphoglyceric acid as the first product of photosynthetic carboxylation via the Calvin cycle (Calvin, 1989).

Most C-4 species are monocots (e.g. grasses and sedges), but the C-4 mechanism also occurs in dicots. All gymnosperms, bryophytes, and algae and most pteridophytes are C-3 plants, as are nearly all trees and shrubs. An interesting feature of C-4 photosynthesising plants is that they appear to be particularly suited to conditions of high light and warm temperatures. It has been shown that C-4 plants can fix CO_2 in the chloroplasts of leaf mesophyll cells (in a process that has a high energy requirement) to produce the 4C organic acids, which are then transported to the bundle sheath cells (which have thicker walls, many chloroplasts, mitochondria and other organelles, and smaller central vacuoles; Frederick & Newcomb, 1971) where the transported 4C acids are decarboxylated and the released CO_2 is photosynthetically fixed via the Calvin C-3 mechanism which is located in the bundle sheath chloroplasts with (in sugar cane) sucrose as the final product. The 3C residual fragment of the decarboxylated 4C acid (pyruvate) is then transported back to the mesophyll cells to repeat the cycle.

The overall effect in the cane crop is the utilization of the abundant energy available in the tropical environment to drive the energy-expensive C-4 mechanism to concentrate CO_2 at the site of its fixation by the C-3 Calvin cycle carboxylating enzyme (ribulose bisphosphate carboxylase, Rubisco), thus overcoming the low affinity of this enzyme for CO_2 fixation. In spite of the apparent high energy requirement of the C-4 mechanism, C-4 plants (and especially sugar cane) almost always show more rapid rates of photosynthesis per unit leaf surface area than do C-3 plants when both are exposed to high light levels and warm temperatures at ambient CO_2 levels. It is believed that C-4 plants are adapted to and evolved from C-3 species in regions of periodic drought, such as tropical savannas, and that when temperatures reach 25–35°C and irradiance levels are high, C-4 plants are about twice as efficient as C-3 plants in terms of dry matter production.

Sugar Cane under Elevated CO_2 Concentrations

Because of its C_4 photosynthetic mechanism, it has been generally assumed that sugar cane photosynthesis is CO_2 saturated at ambient atmospheric CO_2 and hence will not show any increased photosynthesis at the elevated CO_2 levels associated with global climate change. It has, however, been shown (De Souza et al., 2008) that sugar cane grown for 50 weeks at elevated (720 ppm) CO_2 concentrations in open top chambers showed an increase of ~30% in photosynthetic production and an increase of ~17% in plant height compared with plants grown at ~370 ppm. There was also an accompanying increase of ~40% in biomass. Plants grown at the elevated CO_2 concentrations also had ~37% lower stomatal conductance and ~32% lower rate of transpiration giving them a higher water use efficiency of ~62%. Significantly, industrial productivity analysis showed an increase of ~29% in sucrose content from the crop harvested from elevated CO_2 chambers. The reasons for these responses to high CO_2 exposure are complex and may involve various components of the photosynthetic productivity mechanism including water-use efficiency.

Ethanol from Sugar Cane

Of all the crops that have been proposed as possible sources of biofuel, sugar cane is one of the most favoured because the agricultural technology and processing methods are well developed and the crop can be harvested over a protracted season. The potential for a northern Australian ethanol industry based on sugar cane has been addressed by, among others, the Australian Society of Sugar Cane Technologists (O'Hara, 2010). The assessment assumed a continuing annual production of 30–35×10^6 t sugar cane from the existing approximately 400,000 ha under cane – mostly in tropical Queensland, although considerable expansion into parts of Western Australia and the Northern Territory would be possible, subject to a favourable business environment.

The technology would be based on yeast fermentation of the sugar cane molasses:

$$C_{12}H_{22}O_{11}(\text{sucrose}) + H_2O \rightarrow 2\ C_6H_{12}O_6(\text{glucose})$$

$$C_6H_{12}O_6 \rightarrow 2\ C_2H_5OH\ (\text{ethanol}) + 2\ CO_2$$

Because of the production of CO_2 as a by-product of fermentation, the maximum theoretical energy yield would be less than 50%. On a moderate scenario of ethanol production, it was calculated that if ~30% of Australia's current production of sugar cane juice was diverted from crystal sugar production to ethanol production (and if there was a contribution from trash collection and cellulosic ethanol production), then the yield would be equivalent to more than half of Queensland's annual consumption of automotive gasoline.

There has, however, been no substantial move towards ethanol production from cane in Australia, even though the industry has, since the mid-1990s, been overtaken by the Brazilian sugar industry as the world's lowest-cost sugar producer. In that country, sugar cane has been successfully used as feedstock for ethanol production for many decades (De Souza Dias et al., 2015) and most of the mills produce sugar, ethanol and electricity (from bagasse burning). It is anticipated that declining demand for crystal sugar may yet dictate that Australian cane growers and the sugar industry may be forced to follow the same route. There are concerns, however, that substitution of biofuels for part of the fossil-fuel demand will lead to competition for land that would otherwise be used for food production, thus driving up food prices. If biofuel production is restricted to low-productivity land, then the amount of land required would inevitably be more extensive thus increasing the harvesting and processing costs.

Any substantial decline in profitability of the sugar cane industry (due to a predicted decrease in the demand for dietary sugar) will require an extensive program of crop-land rehabilitation. In some areas (especially those under irrigated cane production) this has already become necessary because of increased salinity. At the Kalamia sugar mill region north of Ayr in Queensland, certain areas previously used for irrigated cane production have been abandoned, and a highly saline site was used to test the growth of seedlings of selected alternative tree species (Sun & Dickinson, 1995). The site was divided into low (<0.6 dS m^{-1}), moderate (0.6 to 1.1 dS m^{-1}) and high (>1.1 dS m^{-1}) salinity and planted with seedlings of *Casuarina cunninghamiana* and *Eucalyptus camaldulensis*. Both tree species (after 24–36 months) had high survival rates, but the eucalypt, partly because of its deeper root system, was preferred for reclaiming salt-affected land.

OTHER TROPICAL BROADACRE CROPS

Aside from sugar cane, most of the other broadacre crops that are now grown in the tropical north owe their success to the development of the necessary infrastructure for water storage and crop irrigation. Modelling studies have allowed assessment of the agronomic benefits and long-term risks to sustainable yields of different crops under a range of environmental conditions including those predicted under various climate change scenarios.

COTTON

The story of cotton (*Gossypium hirsutum*) growing in northern Australia is inti-
mately linked with that of the Ord River Irrigation Area (ORIA) scheme located
on the border between the Northern Territory and the north of Western Australia
(see Chapter 2, Figure 2.1). The setbacks suffered in the early years, in particular the cost (and low effectiveness) of crop protection against insect attack and the
problems associated with marketing, have tended to colour subsequent discussion
of large-scale cotton growing in the tropics despite the spectacular successes of this
crop elsewhere in Australia. Research and field trials directed at solving some of the
early problems has continued, with some promising results. It has been proposed, for
example, that switching the growth period to the dry season would avoid the worst of
the pink bollworm (*Plectinophora gossypiella*) damage. The development of transgenic cotton strains has also shown potential to contribute to greater resistance to
various pest attacks (Davies et al., 2009), although some quality and quantity problems remain (short fibre length, discolouration and low yield), suggesting that ORIA-
grown cotton is unlikely, without further research, to be competitive with that grown
in temperate Australia.

A series of paddock-scale trials carried out in the 1990s on a number of private
farms at the ORIA (in collaboration with Agriculture Western Australia) has led
to the development of a comprehensive and sustainable pest management system
that avoids the frequent (and expensive) insecticide spraying program that would
normally be required. It is based on the use of transgenic cotton varieties containing
the INGARD gene (Monsanto) along with an expanded integrated pest management
(IPM) system (Strickland & Annells, 1999). The transgenic cotton variety is one that
has been genetically engineered by the insertion of a Bt (*Bacillus thuringiensis*) gene
which confers resistance to a range of caterpillars. The expanded IPM system avoids
exposing the crop to the highest pest populations by switching to winter cropping.
The 5-month summer break from cropping (except for sugar cane which, anyway, is
virtually pest free) helps to reduce overall pest numbers. The system also makes use
of "trap crops" such as lucerne and niger (an oil-seed crop) to draw some pests away
from the cotton crop.

One of the key findings of these trials was that some of the crops receiving the
IPM treatment, especially those using lucerne as a trap crop, had higher yields and
needed less insecticide spraying, especially if the lucerne was planted in strips. The
trials also confirmed that some of the insect populations associated with the cotton
crop, far from being harmful, were actually either benign or, more interestingly,
beneficial. The wasp *Trichogramma pretiosum*, for example, parasitises the eggs of
one of the most serious pests of the cotton crop (*Helicoverpa* spp.) and is particularly effective when expression of the Bt gene is not strong enough to eliminate the
Helicoverpa. It also helps to prevent overexposure of the *Helicoverpa* to Bt, thus
lowering the selection pressure for resistance to the bacterial transgenes.

Widespread release of the Bt cotton varieties in northern Australia has raised
concerns about the potential for increased "weediness" of this genetically modified
cultivar in non-cropping habitats. To explore this possibility, Eastick and Hearnden
(2006) planted Bt and conventional cotton seed at a series of sites across northern

Australia, comparing their relative invasiveness as indicated by germination, survival and recruitment. There was no observable difference in weediness between the two types of seed, even after 4 years and even after planting at what were presumed to be some high-risk habitats such as irrigation drains. It is clear that there is now greater confidence among growers that production of irrigated cotton is indeed feasible at different sites across northern Australia (Yeates et al., 2013). Establishment of a viable cotton industry would, of course, be subject to the availability of suitable land, adequate water, and government or investment support for the necessary ongoing and locally based research.

The extent to which predicted climate change will affect the yield and quality of cotton production in northern Australia is not clear. Glasshouse trials of cotton grown under enhanced (640 µL L^{-1}) CO_2 concentrations (and 28°C/17°C day/night temperatures) displayed positive growth and physiological benefits compared with plants grown under ambient (400 µL L^{-1}) CO_2 concentrations, provided sufficient water was available (Broughton et al., 2017). At higher (32°C/21°C day/night) temperatures, the effects were negative and were not mitigated by enhanced CO_2 concentrations. There were also indications that climate change might affect tolerance/resistance of some crops (or the weeds) to commonly used herbicides such as glyphosate (Roundup). This would have far-reaching effects since cotton, like many other crops, has been genetically modified to be glyphosate resistant (Fernando et al., 2016).

RICE

Another major Australian crop that has been grown at the ORIA and elsewhere in the tropical north is rice (*Oryza sativa*). It has been cultivated in temperate Australia since the early 1920s, but rather less successfully in the tropical north, although Chinese settlers are known to have cultivated rice in the Darwin area well over 100 years ago. In the 1950s, experimental plots were established under flooded conditions at Humpty Doo (12°33'S, 131°21'E) south-east of Darwin (Chapter 1, Figure 1.1), using imported seed first grown under post-entry quarantine. In north Queensland, rice was grown commercially in the lower Burdekin area for a 25-year period between 1968 and 1993 (Maltby & Barnes, 1986). River floodplains of the northern regions support extensive populations of wild rice including the annual *O. rufipogon* and the perennial *O. auatraliensis,* but, interestingly, the dry Riverina district of New South Wales, Australia's major commercial rice growing area, has no wild rice population.

In northern Australia rice research has been largely devoted to improving yields and reducing fertiliser application to avoid excess nitrate drainage into adjacent waterways. It has been estimated that N fertiliser accounts for about 20% of the total variable cost of rice production in the Burdekin region (Bourne & Norman, 1990). The best production efficiencies were obtained by using short-statured and early maturing rice genotypes (Borrell et al., 1997b), which have now generally replaced the taller and later-maturing genotypes. Among the diseases affecting rice grown in the north is the bacterial leaf blight (*Xanthomonas oryzae*), which has been shown to attack both wild and cultivated rice in the Northern Territory (Aldrick et al., 1973).

One of the major costs of rice cropping in the typically monsoonal climate of the Burdekin Irrigation Area is water supply, particularly during the extended (virtually 6-month duration) rain-free season. Recognising this, Borrell et al. (1997a) compared the yield from rice crops from five different irrigation procedures, namely permanent flooding at sowing, flooding at the three-leaf stage (traditional), flooding prior to panicle initiation and the two unflooded methods of saturated soil culture (SSC) and intermittent irrigation, at weekly intervals. SSC consisted of growing the crop on raised beds (0.2 m high and 1.2 m wide) with water maintained in the furrows (0.3 m wide and 0.1 m below bed surface). It was found that flooding was not essential for high grain yield and quality, since there was no significant difference in yield or quality between SSC and traditional flooded production even though SSC used ~32% less water in both the wet and dry seasons.

Sorghum

Sorghum bicolor has been grown as a crop in tropical Australia for over 50 years both during the wet season (under monsoonal rain-fed conditions) and, since the availability of irrigation, in the dry season as well. Up until the 1970s, the yields were generally poor, but since that time, the introduction of new commercial hybrid varieties and the development of improved tillage methods have led to productivity yields comparable with those obtained with other tropical C-4 cereals.

Muchow and Coates (1986) investigated yield variations of irrigated grain sorghum during the dry season at the Kimberley Research Station (15°39′S, 128°43′E), ORIA, on two soil types (Kununurra Clay and Ord Sandy Loam). The yield comparisons were based on models using data inputs of the amount of photosynthetically active radiation (PAR) intercepted, the efficiency of its use in dry matter production and the proportion of dry matter partitioning to grain. Across the range of sites planted, the best yields came from May sowings (9.5 t ha^{-1}) and the lowest from April sowings (7.4 t ha^{-1}), while the largest variation in grain yield was due to differences in dry matter partitioning. Conversion efficiency (2.4 g MJ^{-1} of intercepted PAR) was higher than the average conversion efficiencies over the entire crop cycle from temperate C-3 cereals growing in a temperate region. Temperature, it was noted, did not affect the conversion efficiency but did affect the duration of crop development.

A series of short-term (4–8 years) experiments and farm demonstrations carried out in the Douglas-Daly and Katherine districts of the Northern Territory, compared no tillage, conventional tillage and reduced tillage for a range of crops, including sorghum (Thiagalingam et al., 1996). With no tillage but with surface application of vegetative mulch, yields were comparable with or higher (especially during the drier years) than those obtained under conventional tillage. In particular, at the Katherine Research Station plots, sorghum sown using no tillage established better, showed better rooting and water utilization, developed more rapidly and produced higher grain yield compared with the same crop grown under conventional tillage. Over a period of 4 years, no-tillage and surface-mulch-treated sorghum outyielded that receiving conventional tillage by an average of 79%. The benefit of surface mulch application was largely due to a reduction of soil temperature and better weed control.

LEGUMES IN SUSTAINABLE AGRICULTURE

Leguminous plants (members of Fabaceae – the pea and bean family) owe their special agricultural importance to their symbiotic associations with a bacterium (*Rhizobium*) which can "fix" atmospheric nitrogen by a process that can be summarised as follows (Salisbury & Ross, 1992):

$$N_2 + 8 \text{ electrons} + 16 \text{ MgATP} + 16 \text{ H}_2O \rightarrow 2 \text{ NH}_3 + \text{H}_2 + 16 \text{ MgADP} + 16 \text{ Pi} + 8 \text{ H}^+$$

nitrogenase enzyme complex

The host plant provides carbohydrate – the source of the reducing power (electrons) to convert N to NH_3 (probably as NH_4^+) in the bacteroids. The NH_4^+ is translocated to the host plant cells as glutamine, glutamic acid, asparagine or other nitrogen-rich compounds (ureides). The legumes, therefore, provide a nitrogen source to increase soil fertility while, at the same time, reducing the requirement for inorganic fertilisers. Their inclusion in crop rotations helps to reduce costs and protects the environment. They are a key component of sustainable tropical crop production and pasture improvement measures (see Atienza & Rubiales, 2017). Legumes include pulses such as beans, chickpeas, peas and lentils, which are harvested for dry grain, as well as a range of other species that provide an important component of cattle forage (see Chapter 5) or as green manure.

Some of the legume species that have been successfully grown as tropical crops in northern Australia are various species of mungbean (*Vigna*). They have been grown in Australia for many decades but in substantial quantity only since the 1970s (Lawn & Russell, 1978) and then centred largely on the Darling Downs in southern Queensland and in parts of New South Wales. Expansion of mungbean production into the tropical north occurred following a survey to identify Australian homoclimates for centres of production in India (Russell, 1976). The survey showed low-level similarity between the climate of Hyderabad (and a number of other Indian centres) with that of Katherine and Darwin (in the Northern Territory) and a number of coastal sites in tropical Queensland.

Five Australian *Vigna* species were reported by Lawn and Watkinson (2002) to occur in a wide range of grassland habitats extending from the coastal foreshore to the central inland desert. One of them, *V. vexillata* (a tuberous legume) was at one time used as a wild food source by Aboriginal communities (Lawn & Cottrell, 1988) and is now much used as a highly palatable forage crop (Damayanti et al., 2010). It is also one of several native legumes whose tubers have been evaluated as a possible source of fermentable feedstock for the production of liquid fuel ethanol. When used in pasture improvement (see later) *Vigna* species are generally selectively grazed and represent a potential germplasm resource for extending the range of genetic diversity available for breeding crop and pasture cultivars (Lawn & Imre, 1991). A new species, *V. angularis* (adzuki bean), recently introduced from China, has been trialled at the Hermitage Research Station as a crop suitable for cultivation under central Queensland conditions (Redden et al., 2012).

Leucaena leucocephala is a forage legume that has long been naturalised in northern Australia. Its potential to support animal production was at first hindered by its toxicity due to its mimosine content (Hutton & Gray, 1959) and by its susceptibility to psyllid (*Heteropsylla cubana*) attack (Bray, 1994). The toxicity problem was solved by the successful transfer of toxin-degrading bacteria from Hawaiian goats to Australian ruminants (Jones & Megarrity, 1986); the second problem was controlled by a combination of breeding for resistance and careful management of the growing conditions.

Leucaena pastures have now been widely adopted in the tropical north, although they were formerly largely restricted to central Queensland. They clearly meet graziers' needs for a highly productive and profitable industry that meets marketing requirements for grass-fed beef of superior quality. Tropical grass pastures, even on fertile soils, are not rich in protein, rarely exceeding 10% of dry matter compared with the 12–13% needed to allow cattle to gain weight quickly and consistently throughout the year. By contrast, *Leucaena* forage contains 20% crude protein, is highly digestible and provides a consistently high-quality diet throughout the year (Shelton & Dalzell, 2007). Importantly, *Leucaena* pastures can fix >75 kg N/ha/year, some of which is cycled to the pasture via animal dung and urine during grazing.

Recognition by graziers that commercial varieties of *Leucaena* have definite weed characteristics has led to the adoption of a voluntary code of practice which, pending the development of sterile varieties or hybrids, offers guidelines for the planning and management of the crop in a way that maximises animal production with minimum weed risk.

LEGUMES IN CROP ROTATIONS

There has been growing concern over recent decades that agricultural production, the world over, is becoming less diverse with the development of cropping systems characterised by low levels of genetic diversity, usually accompanied by increasing use of synthetic chemicals (Liebman & Dyck, 1993). In the tropical north of Australia, as elsewhere in the world, this trend is now being countered by the application of agroecosystem management methods that can restore a measure of crop diversity without compromising (and perhaps, even enhancing) yields and profitability. The introduced crop diversity can be temporal (crop rotation) or spatial (intercropping).

Crop rotation trials are currently under way at Katherine in the Northern Territory in which a legume, *Arachis hypogaea* (peanut), is grown in rotation with maize under different levels of irrigation and different levels of N fertiliser application (Chauhan et al., 2015). Dry season maize is rotated with wet season peanut crop and the yield modelled under imposed climate stressors of 1–2.8°C increases on a dry season baseline of 24.4°C and a wet season baseline of 29.5°C and the data calculated for 2030 and 2050 timeframes. It was predicted that increased temperature would cause a reduction in productivity and total soil organic carbon (SOC) accumulation and greater N losses and greenhouse gas (GHG) emissions. It was nevertheless recommended that the peanut–maize rotation method should be retained because of its overall yield and sustainability advantage in warmer climates, although any limitation of irrigation as a result of climate change would, of course, reduce these advantages.

NOVEL CROPS

The potentially high crop yields available in the more remote tropical regions can best be exploited by the cultivation of novel crops whose high value can justify the high costs of cultivation, harvesting, treatment and marketing of the final product. The crops described next are representative of those trialled in the tropical north with varying degrees of success.

SANDALWOOD

Indian sandalwood (*Santalum album*) has for a number of years been successfully cultivated in large-scale plantations totalling over 2,000 ha of land in the ORIA in Western Australia (see Chapter 2, Figure 2.1). The product, a fragrant oil (with α- and β-santalol as the major components), is extracted from the sandalwood heartwood and used in the manufacture of perfumes, soaps, cosmetics and medicines (Rai, 1990).

Trials at the Frank Wise Institute, Western Australia, have yield oil from 16-year-old trees (planted at a density of 260 trees ha^{-1}) at 0.28 kg per tree (73 kg oil ha^{-1}), with the larger diameter trees yielding the higher heartwood weights and hence the best yield of oil (Brand et al., 2012). Among the factors contributing to the cost and technical complexity of oil production from the sandalwood tree is its parasitic nature (Hewson & George, 1984); it occurs as a root parasitic shrub or small tree and is dependent throughout its life upon its host. Research in the Kununurra area has shown that *Alternanthera nana* (Amaranthaceae) (hairy joyweed) is a successful host plant for pot trials, while *Sesbania formosa* (Fabaceae) (white dragon tree) is a good intermediate host (Radomiljac, 1998; Radomiljac et al., 1999). Some of the critical factors affecting yield of sandalwood oil relate to the choice of suitable host species, the planting density and watering regime, and the length of the rotation period (usually a minimum of 15 years but may be as long as 20–30 years). The technical complexity of sandalwood production, the long span of the yield cycle and the climatic uncertainties are all serious risk factors that the industry has had to contend with. At the time of writing this chapter one of the major sandalwood producers, Quintis, was already in the hands of financial administrators.

Some of Western Australia's sandalwood is harvested sustainably (as a timber crop) under state government supervision from natural stands in state forests and reserves. State and private plantations have also been established (Jones, 2002). As an alternative to the long-rotation oil production from heartwood, some growers have also investigated sandalwood seed as a potential source of oil (and income) (Hettiarachchi et al., 2012). Interestingly, seeds of *Santalum spicatum* (found naturally in most parts of Western Australia) were found to have high oil yields, especially when sourced from the more arid regions. Sandalwood seed oil (main ingredients: 35% ximenynic acid, 52% oleic acid) is now marketed as a cure for arthritis and some other ailments.

KENAF

Kenaf (*Hibiscus cannabinus*, Malvaceae) has long been known worldwide as a source of fibre suitable for the manufacture of paper and other fibre products. It is an annual herbaceous plant with a straight unbranched stem from which long-fibred pulp of high tearing strength can be manufactured either by mechanical or chemi-mechanical processing. Trial plantings in Australia commenced in the 1950s and the first detailed agronomical studies were carried out at the ORIA in 1972 where the frost-free tropical climate and the ready availability of irrigated water permits year-round growth (Muchow, 1981). Fibres from kenaf bark (up to 35% of plant dry weight) are close to the optimum length of 3–4 mm needed for the production of quality paper products, while the core fibres are shorter (~1 mm) but still comparable with those obtained from many timber sources (Done & Wood, 1981).

Kenaf has been successfully grown at many other areas of the north including the Tipperary region of the Northern Territory and the Burdekin Irrigation area in Queensland. Pulp raw material can be produced from kenaf within 6 months of planting compared with the 20 or so years required for the increasingly scarce timber-based pulp. The long-term potential for good kenaf yield in regions of high rainfall variability, both within and among seasons, has been modelled using histori-cal weather data from different regions of the Northern Territory (Carberry et al., 1993). It was found that, generally, the optimum window for sowing was from early November to mid-December. Earlier sowings were only advantageous in those years of good pre-November rainfall. Across the region, the kenaf yields were within the range 800–17,200 kg/ha (mean stem yield 8673 kg/ha) with the best yields coming from the wettest sites (e.g. Adelaide River – 13°06′S).

The economics of a pulp industry based on kenaf in the north would have to consider the cost of establishing pulp milling facilities in remote locations or, alter-natively, transporting the pulp material as bales or briquettes for final processing else-where. That may explain why, as reported in a broadcast by ABC radio (Kelly, 2006), the most economically successful kenaf plantings have occurred on the east coastal region of Queensland (Mackay, 48 ha, and the Burdekin district, 30 ha). The total annual kenaf yield (more than 1,000 t in a good year) is harvested, dried and stored (6–8 weeks), then processed (crushed) and compacted for transportation, much of it to Japan where it is used as an animal bedding product. Another significant factor is that kenaf can be grown in rotation with sugar cane and, moreover, can be harvested using the same equipment as is used for cane harvesting.

AGAVE

The agave plant has long been used worldwide as a source of commercially important products such as sisal (for rope making) and bulk sugar as a fermentation substrate for alcohol production (e.g. tequila from *Agave tequilana*). More recently, agave has been increasingly grown for its yield of biomass, especially under arid conditions and in regions not suitable for more traditional agricultural cropping. Trials carried out in northern Australia, using agave plants imported from Mexico and subjected to tissue-culture propagation, showed that agave could be successfully grown in a wide range of locations, especially in the tropics. At the Kalamia Estate in the Burdekin River Irrigation Area 3,500 plants were planted and monitored for their growth and productivity under a variety of conditions (Holtum et al., 2011).

One of the most significant findings was that *A. angustifolia,* like other species of agave, a crassulacean acid metabolism (CAM) plant (see Chapter 1), could be grown under the pronounced day–night temperature variations of the tropics without significant change in the relative contributions of the CAM and C-3 productivity mechanisms (Holtum & Winter, 2014). In particular, the stable proportion of CO_2 (75–83% of the daily total) fixed at night makes agave well suited for biomass production in a seasonally dry region (when nighttime CO_2 fixation can occur with minimal loss of water by transpiration). Another feature of agave contributing to its efficiency as a biofuel source is its high moisture content (>83% w/w in the leaves) and inversely, its low water requirement (300–800 mm·year^{-1}, compared with 1,500–2,500 mm year^{-1} for sugar cane) (Corbin et al., 2015).

Thus far, the major use of agave in tropical Australia has been as an addition to bagasse (a by-product of sugar milling) used to provide the mill's power requirement. More recent developments include the use of agave feedstock for the production of ethanol to meet the blending levels currently legislated for transport fuels in Australia. A detailed analysis of leaf material of *A. Americana* and *A. tequilana* (Corbin et al., 2015) has shown that 56–60% of the dry weight is in the form of potentially fermentable sugars, over half of it in the soluble fraction. It was further shown that ethanol yield (using two different strains of *Saccharomyces cerevisiae* as fermenting organisms) rival, in yield ha^{-1} yr^{-1}, those obtained from the most successful biofuel feedstock crops. Trials carried out at Ayr in north Queensland (Subedi et al., 2017) have confirmed that local farmers are receptive to the cultivation of agave for bio-ethanol production and optimistic concerning its economic viability. The suitability of existing sugarcane milling infrastructure and technology for transport, farming and processing of the agave crop would be an added advantage, as also would be its possible use as a rotation crop with sugarcane, thus helping to stabilise farmers' income.

TROPICAL FRUIT AND VEGETABLES

While wheat, rice, maize, barley, soybean, cassava and banana make up our major food staples, many other plant products, some of them suitable for cultivation in the tropics, also have the potential to contribute substantially to nutritional enrichment of human diets. Most of them have been introduced to the Australian tropics from other parts of the world and thus face an array of agronomic and horticultural problems of propagation, yield, quality and resistance to pests and diseases. Many of them can only be successfully grown under irrigation and may also require programs of adaptation or genetic improvement. Strictly tropical fruits are largely limited to areas that are completely frost-free – in Australia that amounts to approximately 50% of the total tropical area (Alexander & Possingham, 1984). Among the tropical fruits that have been successfully grown in Australia are paw-paws, avocados, custard apples, persimmon, pineapples, passion fruit, guavas and kiwi fruit as well as a range of field vegetable crops such as asparagus, celery and broccoli.

Some products such as tomato, green beans, pumpkin and capsicum take advantage of early ripening under tropical conditions. Up to 90% of the total supply of tomatoes and capsicum for the Australian market during September and October come from the Bowen-Gumlu region of north Queensland, where harvesting can start as early as May or June.

Production of tropical fruit and vegetables in Australia has for a long time been almost entirely for local consumption, but, over recent years, there has been an increasing demand from neighbouring Asian countries, most notably from Japan, Thailand and South Korea (and from as far away as the Middle East), for products such as navel oranges, table grapes and mangoes, encouraged through Austrade arrangements and, where possible, by Free Trade Agreements. Recent statistics (Molenaar, 2018) show that export of fruit to China, for example, has grown five-fold over a 4-year period, much of it from the tropical north. It is predicted that this trade will continue into the future as Australian growers align their product to the specific tastes of different Asian countries and as more robust quarantine protocols are established.

PROTECTED CROPPING IN THE TROPICS

Generally, growers in tropical Australia have been slow to embrace protected cropping, mainly because the conditions are usually well suited for open field production. Over recent years, however, and especially in anticipation of an increasing frequency of extreme climatic events, there has been an increasing trend towards greater protection of tropical crops from wind, heavy rain and more lengthy periods of high solar radiation. Design features for suitable protective structures have been published (e.g. Carruthers, 2015) and their use has been reported to give improved yields of many high-value crops such as melons (rockmelons and honeydew) and other fruit with short cropping periods, allowing insertion between other crop harvests.

Such protective structures, comprising 60 m long (6 m width and up to 4.5 m height) plastic sheeting tunnels have been successfully used at Giru (south of Townsville). They were additionally fitted with shade cloth, side ventilation and insect exclusion measures. Irrigation cycles were automatically controlled and drainage collected for reuse on other crops such as mango or vegetable crops. The facility has been used to produce tomato, capsicum and cucumber, with future expansion planned to include potatoes and leafy vegetables such as red cabbage.

THE BANANA INDUSTRY IN NORTH QUEENSLAND

Banana (*Musa* sp., Musaceae, Zingiberales) is Australia's major horticultural tropical fruit crop. It comprises some 250 or so banana farming enterprises located across a large area of coastal north Queensland extending from Mossman in the north to Cairns, Innisfail, Tully and as far south as Ingham and extending inland to the Atherton Tablelands (Figure 1.4). More than 95% of Australia's banana crop comes from this region, most of it sold through the supermarket chains. There is also some banana production in the semi-arid region at Kununurra (Western Australia) and at Katherine and near Darwin (Northern Territory).

The Australian banana industry is based almost exclusively on Cavendish culti-vars. Lady Finger varieties (of non-Cavendish cultivars), although preferred by some consumers, are reported by some growers to be more susceptible to *Fusarium* wilt. All edible banana cultivars are essentially derived from two wild seeded species – *Musa acuminata* and *M. balbisiana* – both diploid ($2n = 22$).

From the time of planting, it usually takes approximately 12 months to produce the first bunches (150–200 bananas) with subsequent bunches being produced every 8–10 months thereafter. After harvesting, the parent plant is cut leaving about 2 m of trunk to support the new suckers that grow from the base and produce the next crop. The bunches are covered for protection during harvesting in a green condition and stored in cool conditions (14–16°C). Ripening occurs after return to warmer conditions.

The industry is subject to strict government regulation, especially in relation to disease control and in the movement of plant material. *Fusarium* wilt outbreaks have been reported in Queensland plantations, and it has been found that the most effective measures to control spread of the disease is through the strict imposition of quarantine and exclusion (Pegg et al., 1996). Because banana is a perennial crop, it is generally accepted that the most effective means of controlling *Fusarium* wilt is through breeding for host resistance for which *M. balbisiana,* because of its agro-nomically more robust traits, is an important gene pool for breeding (Simmonds & Shepherd, 1955).

MANGO (MANGIFERA INDICA)

Mango, by production, is the world's second most important tropical fruit crop behind banana (Bally et al., 2009). In Australia, as is the case generally throughout the world, most of the crop is consumed domestically. It was introduced into Australia, from Malaysia and India in the 1800s, but many of today's most favoured mango variet-ies originated in California. In Australia, the preferred cultivar is the Bowen mango (Kensington Pride) and the mango-growing centres are concentrated in Queensland, but smaller centres have been established elsewhere across the north such as Darwin and Katherine in the Northern Territory and Kununurra in the Ord River Irrigation Area in northern Western Australia. There has been a dramatic increase over recent years in mango production in different regions of Western Australia, taking advantage of the early (September) production in the more northerly locations such as Kununurra and Kimberley (Chapter 1, Figure 1.1) and the progressively later peak production (extend-ing to April) in the more southerly regions towards Broome and on towards Carnarvon.

Many mango varieties do not produce fruit true to type from seed; more regular yields are obtained from grafts, especially of Kensington Pride and Florida-coloured varieties. Mangoes are affected by a variety of pests and pathogens, and there is, depending upon the conditions, sometimes considerable post-harvest loss. The most common post-harvest disease is anthracnose, caused by a range of fungal species belonging to the Botryosphaeriaceae (Sakalidis et al., 2011). These diseases may develop on the mango crop in the field but their expression only appears in the ripen-ing fruit after harvesting. The keys to prevention and reduction of these and other related diseases of mango are good orchard management and post-harvest handling and marketing (Johnson, 2008).

ASPARAGUS

Asparagus (*A. officinalis*, Liliaceae), a highly prized perennial vegetable of temperate and sub-tropical regions has also been successfully grown in tropical Australia where its commercial advantage is largely due to its suitability for early harvesting (mid-June to early August) when crops in temperate regions are in winter dormancy and high prices prevail for both domestic and imported asparagus (Bussell et al., 2002). As one of the few perennial vegetable plants, it is well suited to tropical conditions; it grows best in full sunlight and good drainage and has been successfully grown (usually in small <10 ha plantings) in the Ord River area, at Katherine in the Northern Territory and at Mareeba in north Queensland. Special tropical varieties have now been developed (Bussell & Bonin, 1999) and trialled in Western Samoa where the crop was found to be free of fungal disease and insect attack but generally inferior in terms of continuity of yield over successive harvestings.

After planting (from seed or most often from crowns) the plants are allowed to grow for 3 years before the first harvesting of the spears. After harvesting (over 2 to 3 months) the plants are allowed to grow into their mature fern-like foliage which supports growth of the root system to produce the next crop of spears. Attempts to combat foliar disease (such as asparagus rust, *Puccinia asparagi*, or stem blight, *Phomopsis asparagi*) under the high humidity conditions of the tropical north wet season have included the use of trickle irrigation and earlier slashing of mature ferny foliage, although the latter might further reduce the carbohydrate reserves available to the roots and crown, thus reducing spear quality.

TROPICAL PLANTATION FORESTRY

Sustainable harvesting of Australia's tropical productive forests can only be achieved through intensive management of the remaining natural forests allied with the establishment of carefully selected areas of plantation forestry. The latter, in addition to their contribution to meeting the demand for forest products, also take the pressure off the natural forests, their biodiversity and the other environmental services they provide. Large-scale retention of natural forests is difficult to justify on a purely economic basis because of its high cost when set against the profit foregone from other forms of land use such as forest cropping or cattle grazing. That is why natural forests are largely restricted to areas of inaccessible or remote terrain, steep slopes and wetlands.

The timber industry of tropical Australia is centred largely on the Mackay area, on the Atherton Tablelands and along the Cassowary coast (extending from Cardwell, 18°16′S, to Tully, 17°56′S, on the east coast of Queensland) where the major employment emphasis is on the manufacture of wood products. The chief economic woods of the northern forests, as described by Francis (1981), are drawn from the variety of tree families that are represented in those forests (Myrtaceae, Lauraceae, Eleocarpaceae, Rutaceae and Proteaceae) with cabinet woods particularly prized. They include the Silky Oak (*Grevillea robusta*), Queensland maple (*Flindersia brayleana*) and North Queensland walnut bean (*Endiandra palmersonii*). Kauri pine (*Agathis palmersonii* and *A. microstachya*) is a valuable source of softwood. Milky pine (*Alstonia scholaris*) is common in lowland forests as also are acacia cedar (Albizia toona) and black bean (*Castanospermum australe*), the last producing cabinet wood of exceptionally attractive appearance.

Many of Australia's premier cabinet timbers have been traditionally sourced from rainforests of the tropical north. This was effectively brought to an end, despite strong opposition from the Queensland state government and local government authorities, following the nomination, in the 1980s, of the North Queensland Wet Tropics area (Figure 1.2) for inclusion on the World Heritage (WH) listing. Since that time, cabinet timbers have been sourced almost exclusively from freehold forested land not included in the WH-listed area. By way of compensation for the loss to the timber industry following the WH listing, the commonwealth government established a Community Rainforest Restoration Program (CRRP). The program operated across a large area of the north, extending from Townsville to Cooktown, and most of the plantings were on degraded farmlands originally covered by rainforest but cleared for agriculture in the early 20th century. It commenced in 1993 and underwent several changes of emphasis until 2000 when it converted to an advisory role.

SELECTIVE LOGGING

Large areas of the wet tropical rain forests have for many years been subjected to timber extraction by selective logging with consequent disturbance from thinning effects, construction of logging roads and walking tracks, and from soil compaction and consequent increased surface water flow. One such selectively logged site, located in the Mount Spec region (80 km north of Townsville), became part of the WH-listed area in 1988, offering the opportunity for comparison of logged and un-logged plots (Congdon & Herbohn, 1993). The main species extracted were *Acmena resa*, *Syzygium wesa* and *Cardwellia sublimis*, and the differences between the unlogged plots and those logged 25 years previously were still apparent. In particular, surface soil samples from unlogged plots had higher concentrations of N and P compared with those from the gap plots (which were generally dominated by pioneer species). The undisturbed plots had the greater density of trees despite being, in agricultural terms, of low fertility (while the logged plots would be classified as being of very low fertility).

PLANTATION FORESTRY IN NORTHERN AUSTRALIA

Australian forestry has been described as being in transition between an industry based on native forests to one based on plantation forestry. During the first half of the 20th century, substantial tree plantations were established in sub-tropical Queensland where plantations of *Pinus elliotii* and *Araucaria cunninghamii* were established at Beerburrum and at Gympie (Ryan & Shea, 1977). Well before that, however, Queensland was already a provider and distributor of improved *Pinus caribaea* seed to plantations in other parts of the world. By the 1980s, tree plantations had become established in the tropical north of Queensland with different varieties of *Pinus caribaea* generally favoured. Trials at Kuranda (north-west of Cairns) confirmed *P. caribaea* var. *hondurensis* as being superior to other varieties despite its poorer form and windthrow susceptibility (Nickles et al., 1983).

P. caribaea plantations are now distributed widely across tropical Queensland, including at the Byfield State Forest located in coastal lowlands some 70 km north of Rockhampton (Chapter 1, Figure 1.1). There is also continuing research on the use of eucalypts in plantation forests including programs to match the northern Australian provenances of *Eucalyptus camaldulensis* with climatically suitable sites elsewhere in the tropical world. Particular attention has been paid to the problems of establishing exotic pine plantations and to their special requirement for suitable mycorrhizal associations to assist nutrient uptake.

Plantation forestry has as its main aim the profitable production of timber. Equally important is management of the plantation in a sustainable way. In particular, soil productivity must be maintained and nutrient cycles tightly linked. It is now recognised that this can best be achieved by maintaining biological diversity and by allowing or encouraging secondary successional processes that permit recolonisation of the understorey by native vegetation. Studies at the Queensland Forestry Research Institute, Atherton (17°16'S, 145°29'E; see Chapter 1, Figure 1.4) (Keenan et al., 1997) have shown that colonisation of the understorey of monoculture plantations varies between the different crop species being grown. A survey of monocultures of four commonly grown species (the exotic *P. caribaea* and the three native species *A. cunninghamii, Flindersia brayleyana* and *Toona ciliata*) showed that understorey colonisation by native species was most successful (in terms of species number) under *Flindersia,* then under *Araucaria* with least colonisation under the *Pinus* crop. The understorey colonising species were primarily bird dispersed (80–90%) with the rest wind dispersed. Species richness was largely governed by the age of the plantation rather than by distance from the plantation edge.

Against this background, the authors considered different options for managing the understorey component of the plantations. One option was to maintain the original plantation and regard the understorey as transitory, to be destroyed at the time of thinning or clear-felling. Another was to change the land-use objectives from production to protection of biodiversity. This would imply abandoning the plantation and managing the new plant community for biodiversity values. The third option involves thinning the plantation at an intensity sufficient to generate enough revenue to cover the cost of establishing the plantation and then managing the forest for biodiversity values. The fourth and probably most environmentally acceptable option would be to change from an even-aged plantation management system to polycyclic silviculture that yields timber whilst maintaining a high degree of stand diversity. Harvesting of the forest would occur by selection of both plantation trees and new tree species regenerated from beneath the plantation. The choice between these options would depend on the type of forest and on its ecological status with due regard to overriding socio-economic considerations. Over large areas of the tropical north of Australia the preferred option would be one that allows the possibility of rehabilitation of degraded forest at the least possible cost and still allow profitable production.

In an attempt to reverse the decline in species diversity (and hence the sustainability) of tropical rainforests, but still allow a moderate financial return, Community Rainforest Reforestation Programs (CRRP), as mentioned earlier, have initiated mixed-species plantings. Data for three rainforest tree species (*Agathis robusta,*

Araucaria cunninghamii and *Eucalyptus pellita*) have indicated that after 8 years growth, the average tree volume of species planted in a mixture were between 1.2 to 1.6 times greater than that of species grown in monoculture (Erskine, 2004). Increases in both the observed species number and the effective species richness were found to be significantly related to increased levels of productivity as measured by stand basal area or mean individual tree basal area (Erskine et al., 2006). These results confirm earlier findings using tree-growth models based on data (and other inputs) from farmers, which, interestingly, showed that farm plots were more productive than forester-managed plots (Herbohn et al., 1999), perhaps because farms have more fertile land and farmers are likely to spend more time on site.

The potential for successful plantation forestry in tropical Australia rests on the fast rates of growth that can be achieved (under optimal conditions) and on the opportunity for production of high-value, speciality timbers. Against these advantages must be set the extreme climatic conditions with recurring, long-lasting drought periods and the increasingly more frequent cyclonic events. Geddes (2016) has documented the effects of 13 tropical cyclones (with maximum wind speeds in excess of 100 km/hr) that have been experienced at various locations across northern Australia over the period 1998–2015. The nature of the damage caused was recorded as windthrow (tree uprooted), crown damage (leaves and twigs removed and branches torn off), broken trunks, trunk damage (trees still standing but dead) and sand blasting (in coastal plantations).

Various options have been suggested to reduce the severity of cyclone damage. They include the selection of more wind-resistant (but generally slower-growing) species such as African mahogany or teak, the latter also because it tends to lose its leaves readily in high winds. Species such as *P. caribaea* var. *hondurensis* which have needle leaves – and hence display less wind sail – also suffered less damage. Experience from teak plantations in northern Queensland has shown that those located near the coast suffered extensive cyclonic damage, while those further inland (e.g. at Lakeland) were relatively unscathed. Plantations located within reach of an active woodchip market would, however, offer the possibility of salvaging timber even from severely damaged trees.

Data presented at the African Mahogany Plantation Industry Forum (Lindsay, 2011) highlighted *Khaya senegalensis* as one of the most suitable forestry species for the monsoonal tropics of northern Australia. *K. senegalensis* is now grown in over 14,000 ha of plantations managed by African Mahogany Australia in the Northern Territory, yielding high-quality timber logs for the veneer and sawn timber markets.

SILVICULTURE AND ECOSYSTEM VALUES

Silviculture treatment of tropical forests often involves removal of trees of low commercial value with the aim of increasing the growth and timber value of the harvestable trees. Such thinning will inevitably impact on various aspects of the forest's ecology including species diversity and composition. Studies conducted at the Baldy Mountain Forest Reserve on the Atherton Tablelands in north Queensland (17°17'S, 145°24'E; see Chapter 1, Figure 1.4) have assessed the long-term impacts of silvicultural treatments of different intensities upon species dynamics and timber volume

(Hu et al., 2018). Forest plots were established in 1967 and subjected to four treatments: selective logging only (control) and selective logging followed by three different intensity silviculture treatments applied 3 months after logging in 1969 – low-, medium- and high-intensity thinning (by different degrees of ring-barking, brushing or poisoning).

Tree data (dbh ≥10 cm) recording the number of species, species composition and dominance were collected over the period 1967–2015. Before selective logging, all the plots were similar in having 70–75 tree species. After the silviculture treatment (in 1969) the number of species in the plots (in 2015) were 48, 42 and 18 respectively in the low-intensity, medium-intensity and high-intensity treatment. Monitoring of the recovery showed that in the control, low- and medium-density treatment plots the species number recovered to the pre-logging levels within 46 years. Recovery of species number in the high-intensity treatment plots took longer. Long-term studies such as these clearly provide useful information to assist in managing the yield and sustainability of our remaining tropical rain forests.

AUSTRALIA'S CONTRIBUTION TO FORESTRY IN SOUTH-EAST ASIA

Recent estimations have indicated that global deforestation and degradation of tropical forests accounts for approximately 15–17% of all atmospheric CO_2 emissions (Bosetti & Lubowski, 2010). Protecting existing forests from clearing together with reafforestation of cleared land could, therefore, make a significant contribution to reducing CO_2 emissions. For this to be a viable option, however, the communities concerned would have to be compensated for the benefits foregone from development of the land for agriculture or other forms of profitable land use. The economic status of many developing countries, with a shortfall of recurring and capital budgets, would have to be addressed. Ideally, the benefits of protecting forest diversity, as well as other environmental and social assets should be taken into account and costed.

Compared to its vast overall area, the tropical north of Australia has only relatively small remnant patches of what could be properly described as wet tropical rain forest (and mangroves). Its contribution to global reduction of carbon emissions through reducing deforestation or through reafforestation would therefore be limited. Australia does, however, through its overseas aid programs (e.g. Australian Centre for International Agricultural Research, ACIAR), participate with neighbouring south-east Asian countries in "reduced emissions from deforestation and forest degradation" (REDD) programs. Indonesia, for example, has 90×10^6 hectares of tropical forest, much of it farmed by smallholders who might well benefit from the technical and management inputs available from international or bilateral arrangements under REDD financing (ACIAR, 2012a). In Papua New Guinea, 63% of the terrain is covered by forest – most of it held in customary land ownership and therefore available for managed development subject to suitable community and clan ownership arrangements.

Fostering of teak (*Tectona grandis*)-based agroforestry systems in Luang Prabang province, Lao People's Democratic Republic (ACIAR, 2012b), has undoubtedly enhanced the efficiency of the industry through improved tree management systems and through the introduction of improved teak germplasm. Basic research has been conducted aimed at improving production through optimising tree spacing and thinning, incorporation of companion inter-row crops or fodder species, and improved propagation by tissue culture techniques and clone banks. All of these measures have the potential to reduce pressure on the native forests and, when linked to REDD schemes, to contribute to economic, social and environmental benefits to smallholder livelihoods.

Improved plant protection to reduce crop losses due to pests and diseases is another area where overseas aid can make a real difference. The gall wasp (*Leptocybe invasa*), of Australian origin, is a serious threat to eucalypt plantations in the Mekong region of south-east Asia (and indeed worldwide). In view of the limited success, thus far, of conventional pest control methods, an Australian-sponsored project (ACIAR, 2012c) has been launched to devise a biological control program as part of an integrated pest management strategy. This has involved the training of staff from various participating countries in the protocols for collecting gall material (from across Lao PDR, Thailand and Vietnam) and in setting up tests under quarantine conditions, of the effectiveness of introduced parasitoids (e.g. *Selitrichodes neseri*) in combating the gall wasp.

One of the most productive of the more recent collaborative projects was carried out over the period 2009 to 2014 (ACIAR, 2014). It was based in Vietnam and had as its main aim the breeding of *Acacia* species and *Acacia* hybrids to meet the stated objectives of the government of Vietnam for an expanded plantation estate, estimated at the time to cover over 400,000 ha, including over 150,000 ha of clonal *A. mangium* × *A. auriculiformis* hybrids pioneered by Vietnamese scientists. The major collaborating institutions were CSIRO Sustainable Ecosystems, Australia and the Vietnam Forest Science Institute. There was further collaboration with Indonesian foresters, drawing on their long-standing expertise in testing various hybrids and clonal material and in evaluating the properties of the wood product, including its suitability for pulp and paper-making.

The government-funded ACIAR program is part of Australia's commitment to the United Nations Sustainable Development Goals for the region and, as such, has to be justified to the taxpayer. A recent report (Bartlett, 2017) has highlighted a number of areas where aid projects have brought direct benefits to the tropical north of Australia. These include the development of appropriate silviculture regimes (in collaboration with Indonesian foresters) to support the commercial development of Indian sandalwood plantations in the Northern Territory. The development of high-value engineered wood products (such as timber flooring) from small-diameter *Eucalyptus* and *Acacia* timber is the direct outcome of applying advanced timber technology to utilise a resource that would otherwise be wasted.

5 Rangelands and Tropical Pastures

Large areas of northern Australia ($\sim100 \times 10^6$ ha) occur as eucalypt woodlands where the major form of land use for 100 years or more has been beef production. Whilst there has been very little change to the vegetation during much of this time, over recent decades pastoralists have adapted new practices to increase beef production. These include widespread tree clearing to increase grass production, the introduction of legumes and new grass species, fertiliser application, increased stocking levels and the use of feed supplements. All of these changes are likely to strongly impact nutrient cycling, soil fertility and the overall stability of the ecosystem.

An important component of nutrient cycling in the sparse semi-arid grazing lands is litterfall. This has been confirmed by studies carried out at two typical eucalypt woodland sites near Charters Towers (20°05'S, 146°16'E; see Chapter 1, Figure 1.4) (McIvor 2001). The plots occur in a region having an average annual rainfall of 500–600 mm and a light tree canopy cover of ~12% dominated by *Eucalyptus crebra*, *E. drepanophylla* and *Corymbia/Eucalyptus erythrophloia*. The herbaceous layer is dominated by perennial tussock grasses (*Bothriochloa ewartiana*, *Heteropogon contortus* and *Chrysopogon fallax*).

Annual litterfall measured over 3 years at the two sites was between 700 and 1200 kg·ha^{-1} (greatest during September to December, lowest during May to July). Analysis of the chemical composition of the litterfall confirmed that although the concentrations of nitrogen, phosphorus and sulphur were typically low for eucalypts, they represented a substantial contribution to an ecosystem for which these elements are usually limiting. Tree litter was clearly an important pathway for transfer of organic matter and nutrients from vegetation to the soil surface. It was further calculated that at a stocking rate of one animal per 4 hectares approximately, half of the nutrients in the herbage would be consumed (much of it returned to the soil as urine or faeces), the rest would be contained in the tree litter. The latter component was estimated to represent approximately 20–45% of the biomass (and nutrients) being recycled. The litter also contributes to the ground cover thus reducing run-off and soil erosion.

PASTURE IMPROVEMENT

The native grazing pastures of the dry tropics are dominated by tall perennial grasses such as *Heteropogan contortis* (spear grass) and *Bothriochloa bladhii* (Australian bluestem). For improved grazing and increased beef production, such pastures are often oversown with tropical legumes such as *Stylosanthes humilis* (Townsville stylo) or *S. sundaica* (Shaw, 1961). Such sowings are most effective in those regions having average annual rainfalls of over 850 mm. The effectiveness of oversown legumes in improving the nutrient status of native pastures is subject to the availability of soil phosphorus, which may sometimes be so low as to induce P deficiency in the grazing cattle. This is particularly evident during the wet season (Gilbert et al., 1989) when supplementation with superphosphate fertilisation is often required (Gillard, 1979). Prior to sowing, stylo seed pods are subjected to careful heat treatment to ensure optimal wet season germination (Holm, 1973). Heavy grazing and application of superphosphate may, however, lead to drastic changes in the grass species composition with the original perennial grasses largely replaced by annual grasses and broad-leaved weeds. Another perennial grass resistant to heavy grazing (*Urochloa mosambicensis*) may then be introduced to reduce erosion and weed invasion and to foster pasture stability.

Sown grass pastures in the tropics (and in many sub-tropical areas as well) generally suffer what has been described as "run-down", characterised by an initial period of high productivity followed by a decline in productivity often associated with a loss of desirable species (Myers & Robbins, 1991). The most effective way to manage run-down is to increase N supply by including crops or legumes in the rotation, optimising grazing management, establishing beneficial shade trees or changing to stoloniferous grasses. More often than not, however, the most economically sound solution, and one that maintains gain per head, is to reduce the stocking level.

Many northern farmers have widened their search for suitable tropical legumes to include species such as lablab (*Lablab purpureus*), which has been shown to adapt to a range of temperatures and rainfall levels (Murphy & Colucci, 1999). Even under dry conditions, lablab commonly yields 4,000 kg dry matter h^{-1} of above-ground biomass (Clem, 2004). It can improve the N fertility of soils (Whitbread et al., 2005) and has the potential to produce relatively high growth rates in steers (0.8–1.0 kg/head/day) for 70–100 days, so providing good backgrounding and finishing forage for beef production (Clem, 2004). Lablab has been used as the short-term legume phase in pasture rotations with lucerne and butterfly pea in ley pastures in mixed farming systems in northern Australia (Cullen & Hill, 2005). Different lablab cultivars have been successfully introduced to northern Australian pastures and their contribution to addressing the problem of declining soil N, especially in the more marginal cropping soils, is now well recognised.

Dairy pastures require high application of inorganic nitrogen fertiliser to secure good production of forage and high milk yields. In the Australian dairy industry, N application rates of 200–500 kg N ha^{-1} are generally required to ensure year-round production, provided other requirements are non-limiting. Much of the applied N is, however, lost either through leaching and run-off (as NO_3^- generated by nitrification) or as gaseous emissions of nitrous oxide (N_2O) through nitrification and denitrification (Gourley et al., 2012). Such losses add a substantial financial cost to the farmer. Whole farm N-use efficiency can be as low as ~15% and scarcely ever greater than 50%, with significant environmental impacts in the form of pollution of surface water, groundwater, waterways and coastal lagoons (Brodie et al., 2003). Gaseous emissions of various forms of nitrogen will, of course, contribute to anthropogenic build-up of greenhouse gases (GHG).

Experiments conducted at a dairy farm at Ravenshoe (17°36′S, 145°29′E) (Koci & Nelson, 2016) measured the nitrogen-use efficiency of a ryegrass pasture supplied with N fertiliser at rates of approximately 500 kg ha^{-1} year^{-1}. It was reported that 3 months after N application, 27–39% of the applied N had been taken up by the pasture, 23–45% was recovered from the soil and 18–40% had been lost. Emission of N_2O peaked within a day of fertiliser application. The authors conclude that greater efficiency could be obtained by using urea $(NH_2)_2$ CO as a nitrogen fertiliser along with a nitrification inhibitor to reduce the oxidation of urea to NO_3^- in the soil by ammonia-oxidising microorganisms. One such inhibitor is 3,4-dimethylpyrazole phosphate (DMPP) (Rädle & Wissemeier, 2001) and its effect would be to allow retention of N in the root zone. It would also allow lower fertiliser application without loss of productivity and a lower rate of loss of N to the environment.

THE NORTHERN BEEF CATTLE INDUSTRY

The northern beef cattle herd accounts for more than a half of Australia's entire beef cattle population, but there are large areas of the north that are not at all suited to beef production. The climate is harsh, soils and ecosystems are fragile and the wet season confines production to a limited period making it difficult to finish cattle for markets in one production year (Gleeson et al., 2012). In Queensland, the beef cattle industry is focussed primarily on the beef export market, while in the Northern Territory and the north of Western Australia the focus is more on the export of live cattle.

The success of the beef cattle industry in northern Australia owes much to the development of pastures providing good forage under the prevailing seasonal cycles of prolonged dry periods interspersed by much shorter but often intense wet periods. This has, to a considerable extent, been achieved through the introduction of grass species from other parts of the world. One such introduced species is buffel grass (*Pennisetum ciliare* syn *Cenchrus ciliaris*), which was first introduced (from East Africa) to northern Australia in the 1920s. It was sown in Cloncurry (20°42′S, 140°18′E) in 1926 and in Rockhampton (23°22′S, 150°32′E) in 1928.

Buffel grass produces light bristled seeds which are readily dispersed. It also has rhizomes that allow it to spread to form dense swards (Humphries et al., 1991). In some areas of central Australia, for example, its relative abundance has been shown to increase, over a 30-year period, from 5% to more than 80% (Clarke et al., 2005). It is clearly a very competitive grass, with a high resistance to fire, drought and heavy grazing, making it the most widely sown pasture grass in Queensland (McIvor, 2003). Both buffel grass and birdwood grass (*Cenchrus setiger*) are now well established in the Kimberley region of Western Australia where, because of their high protein and phosphorus content and their high digestibility and palatability, they are preferably grazed when they are green and actively growing (W.A. Government, 2017).

Balanced against the undoubted contribution of buffel grass to improving the productivity of northern pastures are some less desirable traits arising from its strongly invasive nature. It has a tendency to form a continuous flammable layer that can support far more extensive and intense fires than comparable native plant communities (Butler & Fairfax, 2003). It can outcompete native grasses with consequent loss of pasture diversity. Buffel-dominated pastures, although initially providing good grazing value, may run down with time (Myers & Robbins, 1991) and then, when the grazers move on to other grasses, the buffel is allowed to seed thus outcompeting the native grasses. Most important, because of its strong weedy characteristics, buffel may invade environmentally sensitive areas.

An important component of the northern beef industry is that driven by the live cattle export trade, which, despite its economic importance to the region, has been shown to be vulnerable to unanticipated disruptions to external markets as well as more predictable but equally disruptive changes to the beef and mutton supply chain. Thus, although the export of live animals from Australia has to comply with the dictates of the Meat and Livestock Industry Act (1997), there have been occasions when the industry has failed to ensure that appropriate standards of animal welfare have been maintained at the receiving abattoirs. In 2011, for example, supply of live cattle to Indonesia was suspended for one month pending the implementation of stricter rules for transport and slaughter of cattle.

This led to the establishment of the Exporter Supply Chain Assurance Scheme (ESCAS) charged with ensuring compliance with international standards for treatment of animals from departure to slaughter. But, as has been reported by Lindsay (2018), breaches of the regulatory chain have continued, many of them supported by video, photographic and eyewitness evidence. They point to serious deficiencies in supervision and compliance with the act and to adherence with standards of animal welfare expected from a modern industry. Yet, the trade has now recovered and expanded to the extent that early in 2018 the first shipment of live cattle to China left the Port of Townsville to satisfy what is predicted to be a growing demand for high-quality beef in a region already receiving chilled and frozen beef from Australia.

Other factors that might influence the supply chain include restrictions on the areas from which cattle can be sourced such as those areas affected by mosquito-borne bluetongue virus. There might also be changes to the supply chains arising from effects of climate changes such as more severe or more prolonged drought periods. Responding to the possibility of such uncertainties, the industry, in its planning and infrastructure investment program, has adopted a modelling framework that links strategic and operational logistics along the supply chain from property to abattoir or port (Higgins et al., 2012). The strategic optimization model allows the industry to select the best paths, volumes and timing of cattle transport from breeding properties to abattoirs, saleyards and ports. It locates the best rest areas (spelling yards or driver rest areas) along the road network. It also takes into consideration driving hours, agistment periods and maximum water depravation time in relation to site capacities, all aimed at minimizing supply chain investment and operation costs (Garcia-Flores et al., 2014).

The future of the live cattle industry in northern Australia is very much tied in with concerns over its well-documented contribution as a major source of greenhouse emissions. It is known to be a significant source of anthropogenic methane (CH_4) emission as is suggested by the predictions published under the National Greenhouse Gas program. CH_4 is second only to CO_2 in its impact as a greenhouse gas. It has a high global warming potential 25 times larger than CO_2, but it has a relatively short atmospheric perturbation lifetime of 12 years (Kemfert & Schill, 2010).

Since much of the CH_4 emission comes from cattle/sheep enteric fermentation, its nature and extent will vary depending on the nature of the diets consumed. Research work at the CSIRO respiration chamber facilities at Rockhampton compared the daily methane production by Brahman (*Bos indicus*) cattle offered either a tropical forage diet or a high-grain diet (Kurihara et al., 1999). The two tropical forage diets offered were a long-chopped Angleton grass (*Dichanthium aristatum*) forage and a long-chopped Rhodes grass (*Chloris gayana*) forage, while the high-grain diet offered comprised long-chopped lucerne (*Medicago sativa*) hay plus a high-grain supplement.

CH_4 production from cattle fed on Rhodes grass was higher than that from cattle fed on both the high-grain diet and the Angleton grass diet. The CH_4 conversion rate (MJ CH_4 produced per 100 MJ gross energy intake) was not significantly different between cattle fed on Angleton and Rhodes grass but was higher than that for cattle fed the high-grain diet. Based on these data, some dietary options were suggested (McCrabb & Hunter, 1999) for reducing methane emissions by northern Australian beef cattle. Cattle that had been "finished" for 2 to 5 months on a grain-fed feedlot diet were estimated to have, over a lifetime, a 34–35% reduction of methane emission per kilogram of saleable beef. There was a clear relationship between live weight gain (LWG) and methane emission. Cattle having high LWG (1 kg day^{-1}) had lower levels of methane production, while those with lower LWG (0.5 kg day^{-1}) had higher methane emissions. Improving both the level and efficiency of ruminant production, it was concluded, would also reduce methane emission.

RANGELANDS AND CLIMATE CHANGE

The rangelands of the tropical north, like those of other regions of Australia have undergone extensive anthropogenic land cover changes. Such changes on what is already an extremely fragile system strongly influenced by southern oscillation cycles and extremes of temperature and rainfall have in their turn significantly impacted on regional and even global climate. The most notable changes have come from past and ongoing land clearing, confirmed by satellite observational data, which has been shown to directly influence climate by changing surface albedo (the reflected fraction of vertically incident radiation) thus influencing solar radiation effects on evaporation, transpiration and ground heat fluxes.

It has been estimated (McAlpine et al., 2007) that Australia-wide, there has been a $1.2 \times 10^6 \, km^2$ (more than 13%) change in land clearing since European settlement, with a growing recent contribution from tropical regions. A comparison of pre-European with modern-day land surface parameters has shown a strong decrease in the vegetation fraction and an increase in albedo over vast areas. An earlier study by Lawrence (2004) modelled the climate impact of Australian land-cover changes and showed that direct changes in surface roughness (by changing woody vegetation cover) causes an increase in strength of surface winds by reducing aerodynamic drag.

These increases in near-surface winds will amplify the shift in northern Australia from a moist north-east tropical air flow to cooler and drier south-east flow. The resulting warmer and drier conditions will impact upon surface and sub-soil moisture, affecting vertical moisture transport processes. It has been reported (McAlpine et al., 2007) that land clearing to suit grazing has already changed the partitioning of available water between run-off and evaporation in a land already stressed and with limited water resources.

As a result of these changes, the rangelands of tropical Australia now occupy a highly transformed semi-arid region, with highly variable rainfall and soils of low fertility (Syktus & McAlpine, 2016). Their economic viability for intensive cropping and even as livestock pastures is, therefore, marginal and likely to deteriorate further with predicted climate changes. In recognition of this, the rate of land clearing has declined since 2000 and is likely to decline further as economic analysis reveals that large areas of these economically marginal agricultural lands would have better economic and environmental return if used for carbon sequestration and other related ecosystem services.

Current opinion (e.g. Grundy et al., 2016) on future land-use options for some of Australia's rangelands has suggested that sustainable economic returns may be more likely to come from carbon plantings, environmental plantings and biofuel cropping rather than from traditional agriculture. This would require intensification of agriculture on the more productive land (to meet the growing demand for food) while restoring the more marginal rangeland to its former savanna woodland state. The latter could be seen as contributing to regional climate protection as well as reducing global greenhouse gas emissions. The economic rationale for such restoration is likely to become stronger as the price of carbon increases and as regeneration of abandoned pastures becomes more widespread.

REGENERATION OF ABANDONED PASTURES

Northern Australia has a long history of extensive forest clearing for agriculture. The Atherton Tablelands (17°21'–17°07'S, 145°42'–145°37'E; see Chapter 1, Figure 1.4), for example, is an extensively modified agricultural landscape covering an area that was originally tropical forest but is now reduced to remnant forest patches covering a total area of little more than about 900 km². Abandoned areas of the tableland pastures generate secondary forest although often arrested at the stage dominated by two *Acacia* tree species (*A. cincinnata* and *A. celsa*) (Yeo & Fensham, 2014). The successful colonisation of abandoned pastures by the *Acacia* species was attributed to the longevity of their seed in a viable form in the soil compared with the seed of other rainforest species which generally remain viable for no more than 6 months. Dominance of the secondary forest by *Acacia* is undoubtedly aided by seed dispersal by frugivores (which feed on the arils) and by the nitrogen-fixing capacity of the coloniser species. Further progress of the succession, however, is strongly dependent upon the availability of mature forest remnant patches in the vicinity to act as a seed source.

Thiaki Creek Nature Reserve (also on the Atherton Tablelands) is currently under investigation to monitor the pattern of recovery from different forms of previous land use (Charles et al., 2017). The reserve comprises 130 ha of largely intact rainforest (previously selectively logged), patches of notophyll vine forest, a small area of highly degraded rainforest regrowth and 51 ha of abandoned pasture. The degraded regrowth forest contains remnant trees from the original forest together with native and non-native early succession species. The remnant forest area is one that was cleared for grazing approximately 60 years ago.

The survey used seed traps and ground-level seed sampling to track the richness and abundance of rainforest seeds arriving at the different forms of abandoned pastures. It was confirmed that recovery in all these areas was strongly dependent upon arrival of seed from the adjacent patches of surviving rainforest. Recruitment from seed (predominantly animal disbursed) was observed at distances sometimes in excess of 10 m into the abandoned pasture areas but at very low frequency. It was concluded that without some restoration assistance, recovery of abandoned pastures into secondary forest would be largely limited by the very low rate of seed dispersal away from forest edges and would depend on the diversity and density of trees in adjacent remnant forests.

A survey conducted as part of the program of the Co-operative Research Centre for Tropical Rainforest Ecology and Management (Catterall & Harrison, 2006) assessed the status of the range of government-sponsored schemes then underway to assist in the restoration of previously cleared tropical and sub-tropical pastures and cropland. It was concluded that despite the high level of community support received, and the high cost of the venture, less than 1% of the total area cleared had by then been satisfactorily restored to anything close to its original rainforest cover. Some of the high cost was attributed to the need to establish a closed tree canopy as rapidly as possible, which required closely spaced plantings over areas much larger than the remnant fragments that had by that time been achieved.

Studies based in the Atherton Tablelands and at lowlands associated with the Barron, Johnstone, Tully-Murray and Herbert River systems covered two major forms of restoration, namely reversal of degradation ("repair") within existing remnant forest patches and revegetation of formerly cleared land. For both forms of restoration planting, the costs involved could scarcely be justified in terms of improved local biodiversity. Possible cost offsets from the establishment of timber plantations would, of course, further limit any likely biodiversity value.

More recent studies from temperate Australia (Lindenmayer et al., 2012) and from other tropical regions of the world (e.g. Lamb, 2014; Crouzeilles et al., 2017) have discussed the relative merits of active restoration as opposed to natural regeneration of woodland ecosystems. Crouzeilles et al. (2017) analysed the findings from 133 tropical forest studies and were able to demonstrate that natural regeneration was superior to active restoration in terms of biodiversity of plants, birds and invertebrates and in terms of five measures of vegetation structure (cover, density, litter, biomass and height). These findings were contrary to the long-held view that natural forest regeneration has limited conservation value and that active restoration should be the default ecological restoration strategy. Clearly the regeneration of abandoned pastures (or previously logged rainforest) is likely to be a more complex and costly process than could possibly have been imagined when the land was originally cleared.

6 Marine Ecosystems

Australia, as an island continent, is dominated in its climate (and hence ecology) and by its location in the Southern Ocean and the strong influences of the major oceanic currents around the coastline. Almost half of Australia's 36,000 km of coastline lies to the north of the Tropic of Capricorn. Along the east coast of northern Queensland, the main influence is from the warm south-flowing East Australian Current, a branch of the westward-flowing south-equatorial current originating in the Coral Sea. Hydrographic data from a series of cruises to determine circulation in the western Coral Sea region immediately adjacent to the Great Barrier Reef (GBR) have shown that the westward flow bifurcates just north of 18° S (14° S during the monsoon season) (Church, 1987). South of the bifurcation point, the flow is south-eastward on the upper continental slope and north-eastward offshore. All the satellite-tracked drifters deployed in that study went aground in the Great Barrier Reef within 30 days of entering the region offshore of the reef.

The warm westward-flowing south equatorial current to the north of Australia branches to form the Leeuwin current, which continues to flow southward along the west coast of Australia. It is a shallow current that lies on top of a north-flowing counter current and, as a result, the continental shelf waters of Western Australia are warmer in winter and cooler in summer than would be typical at such latitudes. Another branch of the south equatorial current contributes to the Pacific–Indian overflow into the Timor Sea to the north and to the Arafura Sea.

OCEANIC CARBON SEQUESTRATION

Carbon, in its many forms, recycles continuously between the oceans and the atmosphere. The rate and extent of these diffusive exchanges are governed by a complex of natural and anthropogenic processes. Atmospheric CO_2 equilibrates rapidly with surface seawater to create a large carbon reservoir from which carbon is transferred by downwelling to the deeper waters where more carbon may be regenerated from breakdown of organic matter. Dissolved CO_2 returns to the atmosphere from surface waters in regions of upwelling so that the whole system acts as a recirculating pump. In the tropical zone, because the solubility of CO_2 is lower in warmer waters than in cold waters, there is generally a net transport of CO_2 from surface waters to the atmosphere. It is this inorganic carbon cycle that is largely responsible for controlling the pH of seawater which is locally in equilibrium.

One of the most immediate effects of global increases in atmospheric CO_2 concentrations is its effect on oceanic pH. Warmer tropical waters are particularly prone to acidification, and it has been suggested that the Southern Ocean may already be saturated with CO_2 (Le Quere et al., 2007). In the warm offshore waters of the Australian tropics, the inorganic carbon cycle and the CO_2 exchanges between the atmosphere and the upper oceanic waters is dominated by the photosynthetic activity of the phytoplankton, in particular by the diatoms and the coccolithophores. The latter group tends to be the more dominant in tropical waters, while the diatoms tend to dominate at the higher latitudes (Cermeño et al., 2008). A number of coccolithophorid species precipitate calcite inside intracellular vesicles, with the potential to contribute to major fluxes of inorganic carbon to the ocean floor (Berry et al., 2002). Acidification greatly influences calcification processes such as those occurring in a range of marine organisms having calcium carbonate skeletons such as crustaceans, molluscs, corals, echinoderms, coccolithophores and calcifying algae such as *Halimeda*, whose rate of production can equal those of corals (Drew, 1983). They will all be affected by increasing acidity, which reduces carbonate availability (Brierley & Kingsford, 2009).

The carbon exchanges associated with the Great Barrier Reef ecosystems have for a long time been a particular concern in view of the observed increases in sea surface temperatures and persistent acidification (Hoegh-Guldberg & Bruno, 2010). The importance of the coral cover in maintaining the integrity and functioning of the reef has focussed primarily on the process of coral calcification. It involves a complex of highly controlled and energetically costly reactions, largely driven by the photosynthetic activity of the microalgal symbionts (zooxanthellae). Coral calcification has both a physiochemical and an organic matrix component. The former is visualised as taking place in an extracellular space beneath calicoblastic cells of the coral ectoderm (Cohen & McConnaughty, 2003), while the latter is mediated via an organic matrix secreted by the coral (Clode & Marshall, 2002). Coral calcification occurs as a series of reactions involving photosynthetic CO_2 fixation in the chloroplasts of the zooxanthellae and release of CO_2 by respiration of the polyps and zooxanthellae (regulated by appropriate transporters). In the presence of calcium, calcification occurs and $CaCO_3$ is deposited in semi-isolated compartments (Levas et al., 2015).

Coral reefs are probably among the world's most vulnerable marine ecosystems. They increasingly face multiple stressors such as elevated sea-surface temperatures, physical damage from increased storm and cyclone activity, algal overgrowth from increased nutrient loading and reduced herbivory in coastal waters as well as increased sedimentation from adjacent land-use practices. They are also subject to repeated attacks from invading species such as the crown-of-thorns starfish and from the effects of overfishing and increased tourism pressure.

SEAGRASS BEDS OF COASTAL AREAS

The first comprehensive survey of seagrass meadows around the coastal waters of tropical north-eastern Queensland was carried out by Coles et al. (1987). A later survey (Lee Long et al., 1996) provided a more detailed account of the seagrasses of the coastal region extending from Dunk Island (17°57'S) in the north to close to Townsville (19°16'S) in the south. This region, extending over more than 25,000 ha of seagrass habitat, was mapped and characterised in terms of species composition and seagrass biomass in relation to various site criteria. A comprehensive review of the status and management of Australia's seagrass habitats was then commissioned by the Fisheries Research and Development Corporation (FRDC), and its report (Butler & Jernakoff, 1999) included an evaluation of tropical seagrass resources and their importance to coastal fisheries and their sustainability. These reviews have provided a baseline for ongoing monitoring of what is an important resource, well known as a nursery habitat for penaeid prawns and a range of fish species as well as being an essential food source for dugong (*Dugong dugon*) and green sea turtles (*Chelonia mydas*) (Lanyon et al., 1989).

Eight seagrass habitat types have been identified, ranging from coastal intertidal to fringing reef and deep sub-tidal habitats around continental islands. The coastal habitats, dominated by *Halophila* and *Halodule* species, are the most prolific along the north-east Queensland coast. *Halophila ovalis* stands out as being able to survive at depths ranging from 0.93 m above mean sea level (MSL) to 15.1 m below MSL. The large sub-tidal seagrass habitats provide an important food source at low tides when the narrow intertidal zone is not accessible to grazers. The distribution pattern of the seagrass meadows is largely influenced by the shelter provided from the effects of waves and currents, with light penetration (water turbidity) and tidal exposure, being additional contributing factors. The positive correlation between ENSO cycles and reproduction and growth of green sea turtles in other regions of the Great Barrier Reef has been attributed to changes in the quantity or quality of their staple seagrass food source (Limpus & Nicholls, 1988). Similar correlations have been drawn between loss of seagrass beds (due to cyclones, floods or various forms of pollution) and local dugong populations (Preen & Marsh, 1995).

Extensive seagrass beds occur across the northern coastal zones wherever the substrate and the local conditions allow. They extend from coastal Queensland and the Torres Strait, around the Gulf of Carpentaria and Groote Eylandt, across the north coast and along the west coast to Exmouth Gulf and Shark Bay (Chapter 1, Figure 1.1). At all these sites the presence of well-established beds is invariably associated with higher fish densities (Robertson & Blaber, 1992) and, in particular, with juvenile stages of tiger prawns.

Cyclones have always been a major destructive force on seagrass meadows along coastal regions of the tropical north. Cyclone Yasi, which caused extensive damage over a wide area of north Queensland in 2011, was reported by Pollard and Greenway (2013) to have resulted in a "complete and catastrophic loss" of seagrasses in Cairns harbour. A later evaluation based on more extensive historical data including aerial photographs and detailed on-site studies concluded that although the seagrass destruction caused in Cairns by Yasi was severe, it was no more than a small part of an ongoing and quite widespread trend (McKenna et al., 2015).

THE NORTHERN FISHERIES INDUSTRY

Northern Australia supports three types of fisheries – commercial, recreational and Indigenous – each type characterised by a diversity of targeted species and fishing methods. The northern fisheries area is generally of low productivity, reflecting the low nutrient levels of coastal waters and the absence of significant nutrient upwelling currents. Its management, under control of state and territory governments and the Commonwealth Government, is largely aimed at preserving the sustainability of commercial fishing and controlling access of recreational and Indigenous fishers to fishing resources through regulation.

THE NORTHERN PRAWN FISHERY

The Northern Prawn fishery, originally largely centred on the Gulf of Carpentaria (Chapter 1, Figure 1.1), but now extending to other areas along the northern coastline, is one of Australia's most valuable commercial fisheries. The main target species are banana prawns (*Penaeus merguiensis*) and tiger prawns (*P. semisulcatus* and *P. esculentus*) harvested by otter trawling. The by-catch includes other prawn species and a range of other genera such as bugs and squid. The industry is under strict Commonwealth Government control by way of gear restrictions, seasonal closures and regulation of vessel size (McPhee, 2008).

Other prawning areas now included in the Northern Prawn Fishery are those at Joseph Bonaparte Gulf offshore of the Western Australia–Northern Territory boundary and the Queensland East Coast Trawl Fishery. The latter, because of its location within the lagoon of the Great Barrier Reef World Heritage Area, is subject to strict constraints of vessel endorsement and the imposition of by-catch reduction devices. This is particularly important given the historically large by-catch from unrestricted otter trawling in the past and the range of species captured including marine turtles, sea snakes and shark (Poiner et al., 1990).

The juvenile prawn population of the Joseph Bonaparte Gulf region (Figure 2.1) and its associated rivers and estuaries is dominated by two species (*Penaeus indicus* and *P. merguiensis*) which are most abundant in mangrove-lined muddy banks and waterways within the mangrove forests. Adult prawn fisheries occur in waters 50–80 m deep as much as 200–300 km from where the juveniles are most abundant, indicating extensive migration during the life cycle (Kenyon et al., 2004).

BARRAMUNDI IN NORTHERN AUSTRALIA

Barramundi (*Lates calcarifer*) is a euryhaline fish species that inhabits a range of freshwater, brackish and marine habitats in northern Australia. The larvae (2.8–5.2 mm) move from coastal and estuarine regions into nearby brackish and freshwater swamps and tidal creeks where they remain for up to 9 months. They then disperse into estuaries, up rivers and along coastal foreshores where they feed upon fish and crustaceans (Russell & Garnett, 1985). Barramundi is a popular and economically important species, particularly targeted by recreational fishers and now much used in aquaculture (see later).

MUD CRAB FISHERIES

Green mud crabs (*Scylla serratus*) are commonly harvested in tidal waters of the Northern Territory using baited pots (McPhee, 2008). The main participants are recreational fishers, and control is through the issuing of licences and the number of pots per licence. Size limits are also imposed and a strict prohibition imposed on the taking of egg-bearing females. Loss of mangrove habitats through the construction of large-scale coastal developments such as the port construction at Port Hedland in Western Australia would greatly affect mud crab populations as they would a wide range of other mangrove-dependent fisheries.

OTHER RECREATIONAL FISHERIES

The dominant form of recreational fishing is angling (hook and line) and, although a licence is required for most fishing activities, there are no strict limits on the number of licences issued. Public consultation and participation in management issues such as those relating to harvesting regulations are often organised through recreational fishing lobby groups or tourism interests.

INDIGENOUS FISHERIES

The Australian government has decreed that people of Aboriginal descent may, under certain conditions, claim a right to fish in accordance with traditions (but not necessarily the fishing methods) established by their ancestors. The dominant species most often harvested by indigenous fishers are mullet, catfish, sea perch, bream, mud crabs and barramundi (Henry & Lyle, 2003). In the tropical north, Indigenous fishers are largely exempt from fisheries regulations, subject to certain local requirements. The continued fishing by indigenous Australians for marine turtles and dugong has, however, attracted criticism from conservationists (e.g. Marsh et al., 1997).

Management of Aboriginal fishing rights in areas of particular sensitivity such as the Great Barrier Reef Marine Park Area presents particular challenges that government authorities have approached through the inclusion of Indigenous representatives in management bodies and through their active participation in decisions relating to these marine resources (Havemann et al., 2005).

TROPICAL AQUACULTURE

The continued worldwide growth of aquaculture over recent years has been, in a sense, a response to overexploitation of natural fish stocks and stagnation of the supply from wild fisheries. There have also been improvements in fisheries management aimed towards biological sustainability and greater regulation of commercial and recreational fishing practices. A recent survey (Jensen et al., 2014) concluded that aquaculture is without doubt the fastest-growing food production industry (with an average global annual growth rate of ~9%), while supplies from wild fisheries have tended to stagnate.

Aquaculture of a range of tropical fish species is dependent upon a ready supply of food source in the form of small invertebrates such as rotifers (e.g. *Brachionus*) or brine shrimp (*Artemia* sp.) before the developing fish can be weaned on to artificial diets. The small invertebrates in their turn must be raised as extensive brood stocks for which microalgae provide the basic food source. For tropical aquaculture, it might be expected that the most suitable microalgal food source would be microalgal species isolated from tropical waters. That this is indeed so has been confirmed by Luong-Van et al. (1999). A number of microalgal species were isolated from coastal waters of various localities in north Queensland and the Northern Territory and grown as batch unialgal cultures before testing for their suitability as a food source for *Artemia salina*. At least seven microalgal species were found to be capable of supporting high growth rates of *Artemia*, the most successful being species of

Cryptomonas, *Nephroselmis* and *Chaetoceros*. Their suitability was related to their fatty acid composition and, significantly, the most suitable microalgae for sustaining growth of *Artemia* also produced *Artemia* of optimal nutritional status (in terms of polyunsaturated fatty acid content).

The possibility of linking aquaculture with groundwater management has been considered for suitable locations such as those engaged in Salt Integration Schemes (SIS), which involve pumping out groundwater to reduce the water table. Other possible water sources would be those pumped during dewatering of mines or coal seam gas drilling projects. For all these sources, however, the quality of the available water would be a major concern. In most cases the pH is low and the ionic composition is deficient in potassium, crucial for fish physiology and development. For these and other reasons, there has been thus far no significant groundwater-based aquaculture enterprises in the tropics matching the successful culture of brine shrimp (*Artemia*) and the microalga *Dunaliella salina* in the southern states (Kolkovski et al., 2010). In the tropical north far more success has been achieved through artificial stocking of rivers and impoundments with introduced fish to replenish existing fisheries or to establish fisheries where none existed before. Such stocking enhances both recreational and commercial fishing, bringing direct and indirect benefit to the community.

BARRAMUNDI

In northern Australia, barramundi (*Lates calcarifer*), an economically important fish species much favoured by recreational fishers, has many characteristics that also make it an ideal subject for aquaculture (Rimmer, 2003). It is a hardy species that tolerates a wide range of cultural and water-quality conditions. Broodstock may be kept in freshwater or seawater, but seawater (28–35‰ salinity) is required for maturation of the gonads. At spawning the sperm and eggs are released and fertilisation occurs externally. The larvae are reared in "clear" or "green" water, the latter containing microalgae such as *Nannochloropsis oculata* or *Tetraselmis* sp., which, in addition to providing a food source, also assist in maintenance of optimal water quality. Live feed is generally preferred over artificial feed, with later stages feeding on rotifers (*Brachionus plicatilis*) and then brine shrimp (*Artemia* sp.). Juvenile barramundi are transferred from the rearing ponds to nursery facilities and then weaned on to artificial diets. Size grading is applied to discourage cannibalism. Grow-out occurs in cages in natural or artificial lined ponds. Barramundi farming in northern Australia is predicted to increase, supported by improvement in growth rates by selective breeding and improved diets.

The wide natural distribution of barramundi in rivers and estuaries between latitudes 10° and 26°S indicates either a wide thermal tolerance or a capacity for adaptation to local temperatures. Barramundi fingerlings isolated from Darwin (~12°S) were grown at different temperatures and compared with fingerlings isolated from Gladstone (23°S) and grown at the same temperature range (Newton et al., 2010). Thermal tolerance (measured as loss of swimming equilibrium) was significantly higher in the tropical population compared with that of the sub-tropical population. The former strain would clearly be better able to survive any temperature-induced stress imposed by climate change.

PRAWN CULTURE

Wild stocks of prawns (e.g. brown tiger prawns, *Penaeus esculentus*) are widely distributed along northern coastal Australia, extending well into the sub-tropics as far south as Shark Bay in the west and well south of Moreton Bay in the east (Keys, 2003). They are euryhaline omnivores with a preference for seagrass habitats. *P. esculentus* has an extended spawning season and the larvae are easy to rear in captivity with a tolerance for high-density culture. The increasing threat from coastal developments to the natural habitats of wild prawn stocks and the dangers of overfishing will lead to a greater emphasis on mass culture under regulated conditions. Prawn culture is likely to expand to meet what appears to be an increasing demand.

One of the favoured prawn species cultured in northern Queensland is the black tiger prawn (*Penaeus monodon*), most of the crop being currently sold on the domestic market. Broodstock (spawners) are collected from coastal waters between Cooktown and Innisfail (see Chapter 1, Figure 1.2) and transferred to culture ponds (approximately 1 ha in size) in rivers and estuaries where they are maintained at salinities of 15–20‰ and temperatures of 25–30°C. Spawning occurs naturally or can be induced by eyestalk ablation to produce larvae that pass through a number of stages.

Farmed prawns, the world over, are susceptible to white spot disease caused by a virus. Affected prawns develop white spots (0.5–2.0 mm diameter) on the inside surface of the shell, the shell becomes loose and the prawns move to the surface and behave erratically. Losses can be substantial but, until recently, Australia was one of the few countries where cultured prawns had remained free of the disease. Then, in 2017, white spot was detected in prawns in the Logan River and in the Moreton Bay region of south-east Queensland. After its eradication from those sites, extensive annual surveillance of over 60 sites throughout Queensland (by the Department of Fisheries and Agricultural Industry Development) has confirmed that the northern prawn farming enterprises have, thus far, remained free of white spot.

GIANT CLAMS

Giant clams (Tridacnidae) are reared at a number of hatcheries in the Asia-Pacific region to supply tropical marine aquaria but mostly to replenish severely depleted wild populations. Overharvesting of *Tridacna gigas* and *T. derasa*, for the meat provided by their adductor muscle and for the shell itself (Lucas, 1994), has led these species to become extinct in areas of Indonesia, the Philippines, Micronesia and southern Japan, prompting the Australian Centre for International Agricultural Research (ACIAR) to sponsor a series of research projects under its foreign aid program (Copland & Lucas, 1988). The ready availability of clean, shallow, warm seawater in the coastal regions of northern Australia has been cited as being suitable for the establishment of a viable clam mariculture industry to supply markets in Taiwan and Japan.

Research to improve the profitability of giant clam culture has continued both within northern Australia and in co-operation with other Asia-Pacific partners such as the Coastal Aquaculture Centre, Solomon Islands. It has been established that nursery growth of giant clams and their symbiotic populations of zooxanthellae is enhanced by supplying a nitrogen fertiliser in the form of ammonium sulphate or ammonium chloride (Grice & Bell, 1999).

REEF ECOSYSTEMS

CORAL BLEACHING

"Bleaching" is the term applied to reef coral when the coral polyps lose their intracellular microalgal symbionts (zooxanthellae) leaving, in the case of hard corals, coral skeletons devoid of their usual colour. It is now well established (e.g. Berkelmans, 2002) that one of the major stressors causing bleaching is sea-surface temperature (SST). Detailed mapping of the spatial pattern of severe bleaching over large areas of the GBR showed correlation (with 73% accuracy) with SST (Berkelmans et al., 2004).

The pattern of recurrence of severe coral bleaching events in the GBR (and in other reefs worldwide) has been highlighted by Veron et al. (2009). They report that severe coral bleaching started to become evident when global atmospheric CO_2 concentrations exceeded ~320 ppm. At CO_2 concentrations of ~340 ppm, sporadic but highly destructive mass bleaching occurred in most reefs worldwide, often associated with El Niño events. At the time of their report (when atmospheric CO_2 concentrations were ~387 ppm) and allowing for a lag of ~10 years between atmospheric CO_2 increases and the SST response, they predicted that most reefs (including the GBR) were already committed to continuing mass bleaching. It was further predicted that instead of the recent historical pattern of a 4- to 7-year return time, mass bleaching might well, in some areas, become an annual event.

The progressive acidification of reef waters, would inevitably affect the growth, not only of coral, but also of other reef components such as the magnesium-calcite-secreting coralline algae. The reef would no longer be a suitable large-scale nursery for fish and other biota, and in the expected continuing increase of atmospheric CO_2 concentration (up to 600 ppm) there would even be a gradual eroding of the geological structures themselves leaving no more than scattered remnants of the reef-occupying refugia of the more resistant species.

Now, at the time of this writing, the massive GBR coral bleaching event of the summer of 2016–17 confirmed that the predictions of 2009 were, sadly, all too accurate. There is evidence of vast acreages of the northern section of the Marine Park (from the tip of Cape York to just north of Lizard Island, 14°40'S, 145°27'E) suffering severe bleaching and coral mortality ranging from as much as 50% of the total cover in places to less than 9% in others, in what was described by the Great Barrier Reef Marine Park Authority (GBRMPA) as the worst bleaching event ever recorded. It was attributed to record-breaking SST values and a strong El Niño. Australian Bureau of Meteorology data confirm that during June 2016, most of the Marine Park had SST values between 1.0 and 2.5°C higher than the June average over the 2002–2011 period, while the maximum SST values for the period February–April 2016 were the highest since records began in 1900.

The massive coral bleaching event of the summer of 2016–17 was most severe in sections of the GBR north of Cairns (~17°S latitude). Further south, in the Keppel

Island region (~23°S latitude) there was, at the same time, some paling of coral colonies but with significantly less severity than was reported for the northern region (Kennedy et al., 2018). A survey of 14 fringing reefs of the Keppel Island archipelago showed that ~21% of the living coral cover (predominantly *Pocillopora* and *Acropora*) was affected. The authors speculate that, on the evidence of past such events, the severity of any bleaching event may be affected by local influences such as recent cyclonic events, run-off from adjacent river catchments and the resilience of the coral species concerned.

Data from an earlier GBR coral-bleaching event – in 1998 and described at the time as the largest ever recorded disturbance of the reef – have indicated that the flow-on effects on other reef biota are not always easily detected (Bellwood et al., 2006). Annual census data of reef fish community structure over a 12-year period spanning the bleaching event showed no detectable effect on the abundance, diversity or species richness of a local cryptobenthic reef fish community. Thus, a single bleaching event enough to result in a 75% mortality in *Acropora* species and almost a total loss of the dominant structure-forming species *Montipora* seemed to be without major effect on reef fishes. Similar reports of apparent resilience (or a limited response) of reef fish to large-scale coral loss have been reported in the case of storm damage (Halford et al., 2004), crown of thorns outbreaks (Williams, 1986; Hart et al., 1996) and coral bleaching (Booth & Beretta, 2002).

The work of Bellwood et al. (2006), dealing as it did with small fish species with high turnover rates, was, however, able to detect marked but cryptic changes in fish community structure. Most of the fish species had maximum longevities of less than one year, so that bleaching effects would impinge upon the entire life history including settlement, recruitment and survival of juveniles and adults. The apparent resilience reported in previous studies may therefore simply reflect the ability of adult fish of long-lived species to withstand disturbance (by moving habitats, using stored reserves or prey switching).

Carbon Fluxes in Corals under Elevated CO₂ and Temperature

It has been estimated that much (perhaps as much as 50%) of anthropogenic CO_2 emissions may be absorbed by the waters of the world's oceans where it forms carbonic acid, thus resulting in acidification of surface waters. In coastal tropical waters such acidification will have particular consequences in those habitats dominated by calcifying organisms. Continued ocean acidification (OA) has been reported to affect coral calcification (Comeau et al., 2013) and a range of other coral functions (including coral bleaching), especially when accompanied by increases in surface water temperature (Jones et al., 1998).

Levas et al. (2015) studied the individual and combined effects of OA and seawater temperature on the coral-mediated fluxes of dissolved organic carbon (DOC) and particulate organic carbon (POC) in two coral species (*Acropora millepora* and *Turbinaria reniformis*) at two levels of pCO_2 (382 and 741 μatm) and two seawater temperatures (26.5 and 31.0°C). At high pCO_2 there was a decrease in DOC flux in both species, with more DOC being retained within the coral tissues. Temperature, on the other hand, did not seem to greatly influence the flux of either form of carbon.

The results were interpreted in terms of the presumed role of mucus (polysaccharide–protein–lipid complex, i.e. POC and DOC) in the feeding and general health of coral (Brown & Bythell, 2005). In healthy coral, carbon from zooxanthellar photosynthesis (and from heterotrophic consumption of plankton) is either secreted as DOC to the water column or translocated to mucus glands for secretion as a surface mucus layer (and sloughed off as POC to yield more DOC). In the stressed coral (high pCO_2) the pathways for secretion of photosynthate or of heterotrophically produced carbon are diminished. It was suggested that under stressful environmental conditions, some corals might alter their DOC and POC fluxes thus conserving organic carbon and perhaps promoting recovery.

The combination of an elevated temperature and oceanic acidification often results in a marked change in the overall composition of the reef biota. Studies at the Lizard Island Research Station (14°40′S, 145°28′E) have recorded that the dominant response is a marked increase in the relative abundance of the cyanobacterium *Lyngbya* in the turf flora (Bender et al., 2014). The data further indicated that the turf flora component of the reef biota would probably adapt well to any future environmental stresses.

BLEACHING IN GIANT CLAMS

The mass bleaching that has now become a recurring feature of a range of coral species also occurs in other reef organisms that exist in symbiotic association with microalgal zooxanthellae (*Symbiodinium microadriaticum*). Giant clams (*Tridacna gigas*) have their symbiotic microalgae located within a branched tubular system within the siphonal mantle tissue (Norton et al., 1992). Although giant clams have a filtration system of feeding, most of their nutrition is derived from their photosynthetic symbiotic microalgae (Streamer et al., 1988; Klumpp et al., 1992). Loss of these microalgae as a result of various forms of stress (Grice, 1999) causes bleaching such as that observed at various locations of the GBR in the Townsville region.

Laboratory studies (Buck et al., 2002) have confirmed that exposure of *T. gigas* collected from Orpheus Island and Nelly Bay sites near Townsville to prolonged exposure to high light intensities resulted in a significant decrease in the number of zooxanthellae per unit area of mantle tissue (from 19.8 ± 0.8 to 0.2 ± 0.2 ($\times$ 10^7 cm^2). The average cell size of the retained zooxanthellae was also reduced. The most marked bleaching responses were evident only in clams exposed to a combination of ultraviolet (UV) light and high temperature. Another observed effect accompanying bleaching was the release of ammonium (NH_4^+) into the water column and blocking of ammonium uptake into clam tissue. Despite these obvious bleaching effects, it would appear that giant clams might be somewhat more robust than coral in their response to stress-induced bleaching.

CROWN-OF-THORNS STARFISH

Other than coral bleaching, one of the most serious threats to the GBR has been that due to infestations of the crown-of-thorns starfish, *Acanthaster planci* (Aseroidea). Since the early 1960s – and the early reports by Endean (1969) – there have been recurring waves of crown-of-thorns infestations, each wave starting in the Cairns/ Green Island region and then progressing south. The *A. planci* adults are voracious feeders on the soft tissues of a range of hard coral species. They are prehensile and have a typical asteroid morphology with a pliable central disc bearing multiple (up to 15) arms. The upper surface is covered by sharp, toxic spines and the lower surface by sucking tube feet. During feeding, the eversible stomach is spread over the hard-coral surface to provide a high ratio of stomach surface to biomass. Digestion of the soft coral tissue occurs when the everted stomach surface spreads over the coral surface and excreted enzymes digest the soft coral tissue (Birkeland & Lucas, 1990). Lucas (1982), based on his studies of the life cycle of cultured *A. planci*, was able to show that the stage with the greatest potential to influence the number of adult crown-of-thorns is the larval stage. During the life cycle, spawning usually occurs when temperatures reach approximately 28°C and the gametes are shed into the ocean. During early development, the fertilised eggs lead to a gastrula stage followed by various larval stages.

Brodie et al. (2005) noted that plagues of crown-of-thorns (COTS) tend to occur about three years after heavy rainfall and strong terrestrial run-off. They were further able to link such plague outbreaks to increased nutrient run-off leading to increase concentrations of larger phytoplankton (<2 μm diameter) in the inshore water of the central GBR shelf. This, in turn, led to outbreaks of *A. planci* adults probably driven by a massive increase in larval population and their transportation to down-current reefs. These observations explain why successful control of crown-of-thorns outbreaks by attacking the adult starfish has only ever been achieved where the population numbers are small and discrete (Zann & Weaver, 1988). The only long-term solution would be elimination of the source of nutrients that control phytoplankton numbers and hence the larval numbers, larval survival, migration and settlement (Brodie et al., 2005; Lucas, 2013).

Although outbreaks of COTS are undoubtedly a recurring feature of Australia's offshore reefs, there is some evidence that they have become more frequent over recent decades. The extent to which this may be due to climate change is unclear. Models describing the trophic interactions between juvenile and adult COTS and two types of coral (fast and slow growing) have evaluated the effects of predation by large fish on adult COTS and predation by benthic invertebrates on juvenile COTS compared with manual removal of adult COTS. The results confirmed the effectiveness of invertebrate predation in reducing juvenile COTS numbers and the relative ineffectiveness of manual removal of adults (Morello et al., 2014).

NINGALOO MARINE PARK, WESTERN AUSTRALIA

Ningaloo Reef, which extends for about 300 km along the coast of Western Australia (21°50′S to 23°30′S) is Australia's largest fringing reef and is situated in the World Heritage–listed Ningaloo Marine Park. It is well known as a favoured site for the aggregation of whale sharks (*Rhincodon typus*) as part of their migratory behaviour in response to different climatic and oceanographic processes. Data from shark abundance records (Wilson et al., 2001) showed moderate correlation with the Southern Oscillation Index but only weak correlation with Leeuwin Current and sea surface data. The available data suggests that this pattern of aggregation is driven by a seasonal and localised abundance of the preferred whale-shark food (predominantly crustaceans and baitfish).

The Ningaloo reefs, like those of the eastern coastal waters, have also suffered extensive coral bleaching events over recent years. The significant bleaching event of the 2010–11 summer season was reported by Depczynski et al. (2013) to have resulted in a 79–92% decline in coral cover over certain areas of reef. In the worst affected areas, less than 6% of the coral colonies were deemed to be still living after an 8-month period of observation. The coral damage was attributed to an exceptionally strong La Niña Southern Oscillation effect pushing warm currents from Indonesian waters southward along the Western Australian coastline. The most seriously affected coral species were the dominant *Acropora* and *Montipora* assemblages (with very low survival among the latter), while the more massive coral species fared reasonably well. Of the two most seriously affected species, the only survivors were among the <10 cm size tips.

Recovery from such extensive reef damage has been shown to be dependent, to a large extent, on the level of connectivity between adjacent reefs (West & Salm, 2003) and on the absence of genetic reproductive barriers. For *Pocillopora damicornis* coral colonies growing along the Ningaloo Reef, good recovery from environmental perturbations was attributed to a high dispersal potential of sexually produced propagules in the most common lineage with positive spatial autocorrelation detected over distances up to 60 km (Thomas et al., 2014).

In 2011, the Ningaloo Marine Park and the adjacent Cape Range National Park were together added to the World Heritage Register, a clear indication of community concerns at the disturbingly close oil and gas tenements of the Pilbara region and the already massive industrial developments less than 100 km to the north (including those at Onslow and at Barrow Island). Exmouth Gulf, described as one of the last intact arid-zone estuaries left in the world, is already earmarked for massive oil and gas-related development despite its well-recognised importance as a refuge and feeder system for nearby ecosystems including the Ningaloo Reef.

GREAT BARRIER REEF AND TOURISM

The extreme vulnerability of the GBR to global climate-related changes has been well documented (e.g. Anthony et al., 2011) as having potentially strong impacts on coral health, coral growth and on a range of coral-related ecosystems. These effects have been a particular concern of the Great Barrier Reef Marine Park Authority, the federal agency tasked with managing the long-term preservation of the reef and its environmental values. Australia also has obligations under the World Heritage Convention to safeguard the future of the GBR and to oversee appropriate conservation and management measures. Among the bodies with a particular role to play in preserving the environmental values of the GBR are the tourist operators. It is they who interact most closely with one of the major user groups, the visitors to the reef, and it is they who have most to lose from a deterioration of the reef's environmental value.

A recent survey conducted by Goldberg et al. (2018), based on interviews with 19 tourism operators in the Whitsundays and Cairns region, confirmed operators' concerns at the threat to the reef environment and a recognition of their responsibility for enforcing minimum standards of environmental behaviour, perhaps by way of a program of eco-certification. Similar findings have come from an earlier survey (Turton et al., 2010) carried out at a number of Australian climate-change hotspots (including the GBR) where the recommendation was for the establishment of a self-assessment toolkit for each region to identify adaptive approaches for predicted 2020, 2050 and 2070 timescales.

7 Mining and Mine-Site Rehabilitation

Although mining, by its very nature, can never be described as a sustainable form of land use, it is generally accepted these days that the mining industry, in addition to providing safe employment for its workforce and support for community development and infrastructure, has a clear obligation to minimise, as far as is possible, its impact on the environment. The latter requirement applies with particular force to the tropical north where most of the mining activity is located in arid or semi-arid country where revegetation (the key component of mine-site rehabilitation) can be a particular problem and where we already have numerous examples of previously worked and abandoned mine sites, many only partially rehabilitated. Former open-cut mines, because of the high cost of backfilling or the problems associated with dewatering (McCullough & Lund, 2006), invariably leave pit lakes, while old underground workings may leave vast tracts of tailings and waste rock dumps that require processing and ongoing management.

The primary aim of mine-site rehabilitation should be to restore long-term stability to the landscape by minimising surface run-off and sub-soil drainage as quickly as possible. A secondary aim would be a rapid deployment of an overstorey (with optimal biomass to suit the prevailing climate) to provide a foliage protective cover balanced, over time, by a shallow-rooted understorey. Although species native to the area should always be preferred, they may have to be substituted by some more easily established species or those with appropriate eco-physiological attributes. The introduced species should match prevailing conditions and, after an initial establishment period, should not require continuing maintenance in the form of fertilisation, extra watering or replanting.

In terms of land coverage, mining occupies only a relatively small fraction of the north's land area compared, for example, with the grazing industry, but its long-term impact on soils, water and the atmosphere is potentially very high. Some of the north's mines and the environmental problems they present are described below.

THE RANGER URANIUM MINE AT KAKADU

One of the major mining developments in the tropical north of Australia is the Ranger Uranium Mine, located within the World Heritage– and Ramsar Convention–listed Kakadu National Park in the Northern Territory. The mine is owned and operated by Energy Resources of Australia (ERA) and located on land owned by the traditional Mirarr people represented by the Gundjeihmi Aboriginal Corporation (GAC). Mining commenced in 1980 and the Ranger Mine Lease is due to expire in 2021.

The main environmental concern centres on the discharge of wastewater from the mine site to the Magela Creek catchment within the Kakadu National Park. The Ranger mining venture has always operated under a strictly enforced system of regulation and supervision by the Northern Territory and Federal Government overseen by the Office of the Supervising Scientist (Department of the Environment).

The location of the mine has raised the intriguing question as to whether the environmental risk to the nearby sensitive areas is largely from point-source influences from the mine or from more diffuse non-mining landscape-scale influences. This question has been addressed by Bayliss et al. (2012), who used biodiversity of the aquatic ecosystems as the end point measure of environmental risk. Mining risk to the surface water pathway was assessed for key mine-associated solutes (uranium, manganese, magnesium and sulphate). Non-mining landscape risks were assessed for weeds, feral pig damage, unmanaged dry-season fire and saltwater intrusion. The assessments indicated that the non-mining landscape risks, at the time of the study, were several orders of magnitude greater than the risks from mine water contamination. The major ecological risk was from the weed para grass (*Urochloa mutica*), which, at the time, was already extensive and spreading rapidly.

Although the Ranger mine is well managed in terms of environmental impact and work-site monitoring, it will always carry the stigma of its enforced initial establishment against opposition from the traditional owners, the Mirarr people. As described by Graetz (2015a), the Australian government, in approving the mine, had to exempt the Mirarr community from rights granted to the First People under the Aboriginal Land Rights (Northern Territory) Act 1976. Both the Ranger and the nearby – at present "mothballed"– Rio Tinto–Jabiluka venture are reported to have suffered initial systematic indifference to the predicted social and cultural impacts that would flow from these mining developments and the establishment of the associated township of Jabiru. Fortunately, as Graetz (2015b) reports, the historical climate of conflict and mistrust between the traditional owners of the land and the operators of the Ranger mine later gave way to a spirit of greater engagement between the community and the various mining stakeholders. It appears that the mining authorities, thanks in part to the work of the company's community relations team, have now adopted an approach that encompasses human rights and social risk in addition to the usual business/economic risk. Roberts (2008) has given a detailed account of the tortured negotiations between the Northern Land Council (representing the traditional owners and the Oenpelli people) and the government leading up to the final agreement that the tribal lands, after closure of the mine, would be handed back to the government to be run as a National Park for 99 years. In the meantime, the Ranger mine site will be left with a clay-lined 1 km × 1 km earth-wall dam (40 m in height) and other potentially equally unstable holding ponds requiring ongoing management and costly remedial work.

It is interesting to note that, partly as a response to public and government concerns, ERA and Rio Tinto decided in 2005 that the Jabiluka deposit would not be developed unless the traditional owners agreed (Trebeck, 2007). In particular, concerns were expressed by the representatives of Mirarr in relation to proposed surface and groundwater management, specifically relating to the tailings dams. Questions were also raised as to how the financial benefits accruing under any Community Development Agreement would impinge upon the "Care for Country" responsibilities of the Aboriginal community.

MINING AT CORONATION HILL

Coronation Hill is located in the Kakadu Conservation Zone (KCZ), an area of close to 50 ha excised from the Kakadu National Park in the South Alligator River region of the Northern Territory. It was formerly subject to mining leases held by the Coronation Hill Joint Venture Group and mined for its gold, platinum and palladium mineralisation. Mining ceased in 1964 and it was not until the 1980s that renewed interest in resuming mining in the KCZ prompted the Resource Assessment Commission (RAC) to invite further submissions from interested parties and from experts on the environmental significance of the area and its historical and cultural values under the Aboriginal and Torres Strait Islander Heritage Protection Act.

The response to this proposal for renewed mining at Coronation Hill represents an interesting case study of the complexities of trying to balance the economic benefits of mining against any likely impacts on environmental, social or cultural values. The background paper prepared for the government (Cook, 1991) evaluated the benefits to the Australian economy through employment, royalties and infrastructure development (particularly during construction) set against the perceived negative environmental, social and archaeological impacts. Listed among the last of these were Aboriginal rock art sites, scattered artefacts and occupation sites. The report referred specifically to ancestral Bula sites and to the importance of these sites to the Jawoyn people. The land itself was described as being prone to erosion and the soils deemed to be of poor quality. The main source of freshwater, the South Alligator River, was noted to dry out to a series of isolated pools in the dry season making groundwater reserves essential for any significant agricultural development. The region was, however, described as having some potential attraction to tourists seeking a wilderness experience.

In summary, the RAC report concluded that a single mine, properly managed and maintained, would have only a small direct impact on biological resources and little effect on archaeological resources. Although a minority Jawoyn group (mostly younger men) were strongly in favour of mining, the majority, including the nominated custodians, were unequivocally opposed. The RAC even attempted to evaluate the benefits of not mining Coronation Hill but in the end did no more than offer the government a series of options. In the end, the Commonwealth Government, despite considerable criticism, made what could only be described as a political decision informed, no doubt, by the environmental and social climate of the time, and decided that mining should not proceed.

The Coronation Hill mining site was, however, later used experimentally to test the effects of leakage of residual pollutants from the previously worked mine on the downstream environment. Water was gravity-fed from the mine adit to create a point-source pollution downstream (Faith et al., 1995), with regular monitoring of the macroinvertebrate communities both before and after disturbance at a control and impact site as part of a multivariate analytical process. This allowed identification of those taxa showing poor discrimination between pristine and disturbed locations and thus, over a 5-year period, provided a statistical method of monitoring relatively small changes in community dissimilarities. This experiment and others of a similar nature contributed significantly to the establishment of monitoring programs at other mining operations (such as the Ranger Uranium Mine) elsewhere in the Alligator Rivers Region (Humphrey et al., 1998).

MINING IN THE PILBARA

The Pilbara (see Chapter 1, Figure 1.1) is a region of north-western Australia, noted for its richness of mineral resources, with active and pending mining tenements covering a large proportion of the total area but largely confined to ironstone ranges and greenstone belts (Van Vreeswyk et al., 2004). The region produces more than 90% of Australia's iron ore. Such extensive mining activity in a semi-arid to arid landscape presents a particular challenge for rehabilitation and restoration of the environment during and after mining. Despite its exposure to periods of severe seasonal drought, the region supports a large diversity of plant species (Van Vreeswyk, 2004) spread over a wide variety of soil types and conditions. Any post-mining rehabilitation program must, therefore, be preceded by trial plantings to establish the tolerance of different re-colonising plant species to the necessarily much altered restoration sites.

Such trials were conducted by Lamoureux et al. (2016) using nine different *Acacia* species grown in typical mine-site restoration substrates ranging in soil types from alluvial (fine-textured) to sandy and rocky soils. It was found that soil type was a poor predictor of drought tolerance compared, for example, with some other specific traits displayed by different species, indicating that successful mine-site restoration requires careful matching of the preference of the species concerned with the conditions provided by the restored site.

THE RUM JUNGLE MINE

Rum Jungle is the site of a former uranium–copper mine some 105 km south of Darwin, Northern Territory, which was actively mined for export of uranium from the 1950s to 1971. By 1980 it was clear that, since little or no rehabilitation had been carried out when the mine ceased production, the site had become so contaminated as to be not suitable for handing back to the Kungarakan and Warai traditional owners as part of the Finniss River land claim. In particular, the receiving waters of the east branch of the river, which had already been diverted for 1 km of its length to accommodate one of the two mining pits, had become adversely impacted by acid rock drainage from the site. During the early years of the mine's operation, mine tailings were discharged as liquid effluent (pH 1.5) to adjacent low-lying areas and inevitably into the Finniss catchment, affecting approximately 100 km^2 of the floodplain. A survey conducted in 1971 detected no biota for 15 km downstream of the mine. The major problems were attributed to the pyritic nature of the overburden and the environmental impacts resulting from pyritic oxidation (Allen, 1986).

Mine site remediation began in 1982 with the placement of capping rock material over the sulphidic waste dumps to reduce water infiltration. Impacted and unimpacted stretches of the river were monitored for water quality and for annual cycles of contaminant loads (Jeffree et al., 2001). Samples from impacted sites of the river taken before remediation showed reduced diversity and abundance of fish species compared with unimpacted sites. Post-remedial samples from the impacted stretch of the river showed recovery of the fish community with very little difference between them and samples taken from the unimpacted sites. In particular, although considerable contaminant loads (Cu, Zn, Mn and sulphate) were still being delivered to the impacted stretch of river, significantly, there were none of the extensive fish kills that were recorded prior to remediation, especially during the first flushes at the beginning of the wet season.

Remedial work has continued at the site and concerns continue to be raised at the ineffectiveness of the capping material (Taylor et al., 2003) and the status of the remaining pit lakes (Boland, 2008). A more recent re-evaluation of the site (Mudd & Patterson, 2010) has concluded that the rehabilitation process, in particular the design and performance of the engineered soil covers, has not stood the test of time. The Finniss River is still impacted by acid drainage and there are new concerns at possible effects on adjacent groundwater.

MINING IN THE MOUNT ISA REGION

The discovery in early 1923 of mineralised outcrops by the banks of the Leichhardt River in north-western Queensland was the starting point of what was later to become the Industrial complex of Mount Isa Mines (20°44'S, 139°28'E) (Chapter 1, Figure 1.1). The mine quickly became one of the most productive mines in the world (and Australia's largest) with some of the ore bodies containing over 70% lead–silver, later expanded to include copper and zinc. The region around Mount Isa was originally occupied by the Kalkadoon tribe (Horton, 1976) who were able to survive the semi-arid conditions through their judicious use of the few naturally occurring waterholes. Much later (in 1994) the descendants of the Kalkadoon tribe filed a native title claim to over 40,000 km^2 of the Mount Isa region. Native title was eventually granted in 2011 and this allowed the Kalkadoon people to negotiate with pastoralists for access to their traditional lands for hunting and ceremonial purposes.

In the meantime, the rapid development of the mine and the establishment of the township of Mount Isa had been supported by extensive infrastructure development in the region, including road and rail construction and the creation of a number of water impoundments (e.g. Lake Moondarra, Lake Julius and a series of other smaller lakes) along the Leichhardt River and some of its tributaries. The latter development guaranteed continuity of water supply to a region previously lacking any major permanent waterbodies (Finlayson et al., 1984a). It also provided opportunities for water-related recreational activities for a township located in a region in which the annual rainfall can be as low as ~160 mm (but alternating with occasional years of over 800 mm rainfall), with no access to coastal amenities. It also allowed the establishment of an active program of fish stocking, originally sourced from rivers in the region and, more recently using fish raised in hatcheries of the Mount Isa Fish Stocking Group funded by local business interests and government grants.

The hydrobiological implications of the introduction of such an altered aquatic eco-system in such a semi-arid region have been described elsewhere (Griffiths, 2016).

Mine performance (2005 figures reported by the then mine owners Xstrata) delivers annually 4.78×10^6 t ore of which 4.4×10^6 t is treated on site, yielding 231,000 t zinc concentrate, 160,000 t lead-in-bullion and 11.36×10^6 oz of silver in crude lead. The copper stream (accessing separate ore bodies from those providing the lead–zinc) delivers 6.2×10^6 t (2008 figures) of copper from two underground workings (Xstrata, 2010). The environmental impacts of this mining activity, since the early days of production, have received considerable attention supported by ongoing expenditure and research.

Air pollution from the smelter stacks has been an ever-present environmental issue requiring constant monitoring to the extent that smelter operations have been routinely restricted to periods of low risk to the community and completely shut down during periods of westerly winds. More recent use of improved scrubbing technology and installation of an acid plant for converting SO_2 into sulphuric acid have significantly reduced the impact of the mining activity on the workforce, the community and on the surrounding natural environment.

Recognising the semi-arid nature of the region and the unavoidable impact of mining on the groundwater system, the owners (originally Mount Isa Mines, then Xstrata and, since 2013, Glencore) have attempted to maximise use of surface water, stored stormwater and recycled water for mine processing. Mine dewatering is estimated to contribute an average of 7 ML per day into the recycling system and the mine aims to restrict freshwater use to 400 L per tonne of ore milled at the zinc-lead concentrator and 300 L per tonne at the copper concentrator (Xstrata, 2010). Managing mine-site water use involves, on the one hand, husbanding what is a scarce resource during the prolonged dry season, and on the other, coping with the occasional periods of torrential tropical storms, and the threat they pose to off-site, downstream spillage. An important component of the off-site impact of mining is the extensive system of tailings dams, at present numbering eight or more large impoundments constructed by pumping the tailings slurry into extensive settling ponds from which the water is decanted and recycled as process water for use on-site.

As some mining activities come under increasing economic pressure – due to market forces or escalating costs – attention has inevitably been directed towards winding up and rehabilitation of the Mount Isa Mines site, as required by the terms of the lease. The mine has, since the early days, maintained an ongoing program of rehabilitation and revegetation of the tailings dams and, because of its proximity to the township, has always been conscious of the need to reduce to a minimum any off-site impacts such as those from blasting, dust creation from unsealed roads and air-quality problems. As recently as 2017, the copper operations came under particular threat due to, according to the owners, the increasing cost of energy affecting both the smelting operations at Mount Isa and the refinery at Townsville. More recent predictions have indicated that copper smelting may have to end in 2022.

A number of smaller mine sites in the Mount Isa region have already been successfully returned to a state approximating that of the surrounding country. For example, the Mary Kathleen uranium mine, which was formally closed in 1982 after yielding, over the years, a total of 9,000 t of uranium, was completely dismantled and rehabilitated to the extent of being deemed suitable for grazing (World Nuclear Association Publication, 2009). However, a later study, almost 25 years after closure of the mine, reported seepage of metals, metalloids and radionucleotides from the former mill tailings repository (Lottermoser & Ashley, 2005). The seepage occurred from the base of the tailings dam retaining wall at a rate of 0.5 L s^{-1}; it was acidic (pH 5.5), metal-rich and radioactive. It flowed into the drainage system and, although the seepage rate was low (~5 Kg U per year), surface waters downstream of the tailings impoundment were observed to have total dissolved solids (TDS), U and SO$_4$ concentrations exceeding Australian water-quality guidelines for livestock drinking water (especially during periods of low flow). Clearly, maintenance and management of such capped and rehabilitated tailings dams is an essential ongoing requirement. Revegetation, normally an important component of successful rehabilitation, may sometimes present an additional hazard as was shown by another study (Lottermoser, 2011) which showed that colonisation of the Mary Kathleen site by the alien plant species *Calotropis procera* produced plant biomass (leaves, stems and flowers) of potential dietary risk of toxicity to grazing animals.

Another mine site, the Ernest Henry (copper and gold) mine, ~160 km from Mount Isa, has, since 2002, been the subject of a rehabilitation program aimed at returning the site to a stable and safe form of land cover conforming as closely as possible with the pre-mining native grassland (Vickers et al., 2012). Annual monitoring of the vegetation cover has been conducted (since 2007) at rehabilitated sites and at selected reference sites including assessment of vegetation composition, ecological structure and function. The oldest rehabilitated sites were found to have species composition similar to the reference sites, although plant cover and biomass (particularly of native grasses) was lower than in both grazed and un-grazed reference areas. The youngest rehabilitated sites had little established vegetation. It was concluded that the development of a post-mining stable and sustainable grassland is very much influenced by local conditions including soil fertility.

Rehabilitation of the much greater expanse of tailings dams at Mount Isa mines is a work in progress. It has been undertaken through a collaborative program involving the mine and the University of Queensland (Department of Agriculture) and James Cook University to determine the optimum requirements to allow establishment of a stable vegetation cover on successive tailings dams as they become available (Ruschena et al., 1974). Following initial capping of the dried-out tailings using various rock sources, a 1 m depth of dolomitic siltstone (from nearby open-cut mining) was applied along with amendments such as fly ash from the company's coal-fuelled power station and mulching material supplemented by fertiliser inputs (delivered through modified spray equipment). Grasses such as Rhodes grass (*Chloris gayana*), buffel grass (*Cenchrus ciliaris*) and couch (*Cynodon dactylon*) were then sown to assist in the establishment of an organic component. The most successful colonising tree species was *Eucalyptus brevifolia* (snappy gum) already widespread in the area, growing on hillsides and ridges as well as in open flat areas among spinifex bushes (Horton, 1976). Seeding with previously collected native shrub and grass species and planting of seedlings (supported for the first 2 years with trickle irrigation) led to the establishment of a good ground cover dominated by a range of grasses and some shrubs such as kapok bush (*Aerva javanica*), soap wattle (*Acacia holoseri-cea*), whitewood (*Atalaya hemiglauca*) and western box (*Eucalyptus argillacea*) (Lassiere & McCredie, 1983).

BAUXITE MINING AT WEIPA

Aluminium is one of the most abundant (and most used) elements in the world. It and its alloys have numerous industrial uses. It is much sought after in engineering, transport, construction and packing. It occurs naturally as the ore bauxite, deposits of which are particularly plentiful in northern Australia (e.g. at Weipa, Mapoon and Aurukun on the west coast of Cape York Peninsula) (Chapter 1, Figure 1.1). Mining in the area started in the 1950s and '60s and immediately set in train a long series of disputes between the mining companies, traditional Aboriginal land owners, the state government and other bodies such as local Christian missions, as chronicled by Roberts (2008). One of the major issues dominating these discussions is the inevi-tably extremely destructive nature of bauxite mining. It involves stripping away the usually rather thin layer of topsoil and then strip-mining the exposed bauxite-con-taining clay layer, which may be many meters in depth and may extend for hundreds of hectares, yielding approximately 30×10^6 t of bauxite annually.

Rehabilitation and restoration of such land is very costly and technically very difficult. The highest yielding bauxite ores occur beneath tribal lands occupied by the traditional owners who have, understandably, always been strongly critical of the industry. During the early exploration and discovery of the bauxite deposits, some local indigenous people were closely involved but, after closure of the nearby Mapoon mission (in 1963), the Aboriginal people were forcibly removed. The developers (first Comalco, then Rio Tinto) sought to work in partnership with the indigenous people through a number of agreements and trusts to fund sustainable community initiatives to overcome what were identified as barriers to indigenous participation in the min-ing activities and in the wider community (Ross, 1992). Recent Minerals Council of Australia data indicate persistently low participation of Indigenous people in the mine workforce, something that seems likely to continue during the proposed expan-sion of bauxite mining (by Rio Tinto) to land owned by the Wik-Waya people lying between Weipa and Aurukun.

Current legislation requires that before any mining activity or land clearing, there should be a comprehensive survey of the flora and fauna of the area and an assess-ment of any habitats of particular ecological significance. The contribution of local Aboriginal communities to this process, especially in collecting seed for revegeta-tion studies and more generally through their participation with the Department of the Environment and Heritage Protection in site rehabilitation, has been reported in the company's Sustainable Development Report (Rio Tinto Weipa Operations, 2017).

Rehabilitation of land subjected to bauxite mining represents challenges as great, if not greater, than those encountered following strip-mining for coal. Progressive rehabilitation of bauxite-mined areas at Weipa began as early as 1966, first with the aim of establishing commercial pastures, forestry and horticultural crop-ping but since the 1980s with the more realistic aim of establishing some form of native pastures. Trials of various post-mining rehabilitation operations (Schwenke et al., 1999, 2000) met with scant success due to a failure to establish and retain a good organic layer. Previously stockpiled surface soil layers were found to decline rapidly through accelerated oxidation, both during storage and after spreading. Disc ploughing caused even greater loss of organic matter, while incorporation of low-grade bauxite and ironstone had the effect of permanently lowering the rehabilitated soil's productive potential.

Surface soils that were respread either immediately after stripping or after stockpiling for several years were revegetated with *Khaya senegalensis* (African mahogany) and *Brachiaria ceumbens* (para grass) *Stylosanthes* (stylo) mix. Compared with undisturbed native forest plots, they generally had shallower soil layers and more compacted, more gravelly soils. This applied particularly to those spread with stockpiled soil. It was estimated that over the 22-year period of this trial, organic matter had accumulated in the surface soils at rates up to 1.25 t C year^{-1} but apparently without reaching equilibrium level. Nitrogen mineralisation also increased over the time period. It was recommended that, in order to maximise soil organic matter development, post-mining stockpiling of the removed topsoil should be avoided.

For every 3 tonnes of bauxite ore processed, approximately 2 tonnes are left as a caustic residue. The coarser fraction of this residue is used during rehabilitation of the mine site to form outer embankments to contain the finer (red mud) fraction upon which revegetation takes place. Revegetation of the residue sand is, however, very much hindered by its highly alkaline, saline and sodic nature. It also has poor nutrient availability and poor water-holding properties. Various amendments have been used to remedy these deficiencies such as piggery waste, biosolids, compost, green waste and biochar. Another constraint to revegetation of the bauxite residue is its deficiency of the essential element manganese (Mn), which has been addressed by various amendments and through the selection of plant species able to survive conditions of low exchangeable Mn levels (Chitdeshwari et al., 2012).

KIDSTON GOLD MINE

Located some 200 km north of Hughenden (19°51'S, 144°12'E; see Chapter 1, Figure 1.4), the Kidston mine, one of the largest open-cut gold mines in the north, was actively mined between 1921 and 1945 and again from 1984 until its closure in 2001. The mine's water requirements were met from a dam built on the nearby Copperfield River. After closure of the mine, ownership of the Copperfield Dam was handed back to the state and downstream properties continued to receive water by pipeline. Water releases from the impoundment helped to restore downstream waterholes for stock use. An 8-month program of toxicity testing was carried out on stock grazing on rehabilitated tailings residues and on vegetation in the area (Bruce et al., 2003). Analysis of liver, muscle and blood heavy metal accumulation showed that of all the metals tested, only arsenic and zinc were detected and then only at trace levels, while risk assessment analysis indicated no likelihood of any serious metal toxicity or long-term contamination. The results did, however, indicate that a good vegetation cover, through preventing exposure of tailings material, would reduce direct heavy-metal ingestion and hence also reduce the risk of toxicity.

More recent developments at the mine site include the construction of a solar farm with the possibility of a link with the Copperfield Dam (see Figure 1.4) to provide a pumped storage hydroelectricity system. Construction of this system commenced in 2019, with power generation predicted from 2022, providing up to 8 hours supply to the grid.

OPEN CUTS AND BIG PITS

The history of mine-site rehabilitation in Australia includes examples where the mine operators have made serious attempts to minimise the environmental impacts of their activities as well as some where such considerations have received what appears to have been a low priority. Sadly, examples of the latter are often found among some of the country's oldest and most productive mines. One such mine is the Mount Morgan Mine, located some 30 km south of the tropical city of Rockhampton (23°21′S, 150°32′E; see Chapter 1, Figure 1.1). Mining at Mount Morgan commenced in 1882 with the production of copper, gold and silver first as underground mining and later (after flooding of the mine in 1927 to contain an underground fire) by open-cut methods. Over its most productive period, the mine is reported to have yielded 250 t gold, 50 t silver and 360,000 t copper and to have made a major contribution, at the time, to clearing Australia's then national debt.

The legacy of all this productivity was a heavily contaminated mine site and a big pit containing strongly acidic, contaminated water, which, after mining ceased in the early 1980s, required ongoing management to prevent downstream contamination of the Dee River. Management included lime-dosing and on-site evaporation as well as controlled releases from the pit to prevent uncontrolled releases during periods of heavy rain. By 2000, management of the site had passed from the previous caretaker role to one of active rehabilitation (Unger & Laurencourt, 2003) overseen by the Queensland Department of Natural Resources and Mines. There has been little improvement and, as recently as 2013, heavy cyclonic rains caused overflow from the big pit to contaminate the Dee River.

As an alternative to what has thus far been a costly but mostly ineffective program of mine-site rehabilitation, some serious consideration has also been given to linking rehabilitation with what was hoped might be a profitable re-working of mine-site spoils to recover harvestable products. Over recent years, a number of proposals have been submitted, most recently (in 2017) by Carbine Resources who planned to extract copper, gold and pyrite from the residual tailings as well as treatment of big-pit water to produce copper sulphate. The proposal received the approval of the Queensland Heritage Council., but for environmental and other local concerns was later withdrawn.

It is now generally recognised that recovery of the vegetation component at a rehabilitated mine site may not, of itself, be an accurate indicator of complete ecosystem recovery. Studies carried out at three open-cut coal mines (Blair Athol, Newlands and Callide) in the Tropic of Capricorn region of central Queensland (Houston et al., 2018) monitored the recovery of reptile, amphibian and mammal assemblages at a range of rehabilitated sites. Comparisons were made between sites subjected to short- (up to 7 years), mid- (8 to 15 years) and long-term (24 years) rehabilitation. Ground cover of early age rehabilitation was dominated by tussock grasses, which, by mid-age rehabilitation, had declined to give way to saplings and a ground cover of leaf litter. Late-age rehabilitation was characterised by development of an arboreal habitat, although some saplings and small trees had died. It was found that, after 12 years rehabilitation, the reptile species assemblages had approximately two-thirds of the average species richness of nearby reference sites and were assessed as displaying 70% of the average similarity levels between reference forest sites. Species with more specialised foraging habitats (e.g. lizards) took longer to re-colonise than species with more generalist habitat requirements.

COAL-SEAM GAS PRODUCTION

Coal-seam gas (CSG) is gas trapped in coal seams that can be extracted through wells sunk into the seams. Reduction of the groundwater pressure allows the gas to flow towards the production well. The gas originates either from the same biogenic processes that formed the coal or from thermal heating of the buried coal. It is adsorbed within micro-fractures within the coal seam and held there by the overlying pressure including groundwater pressure. In northern Australia, the main centres of CSG production are the Walloon coal measures of the Surat Basin of central Queensland and the Bandanna Formation of the Bowen Basin. They supply gas to the pipeline grid and to local gas-fired power stations. They also make a substantial contribution to Australia's export earnings via a number of liquid natural gas (LNG) compression plants.

The gas, predominantly methane (CH_4) with traces of CO_2, nitrogen gas and ethane (C_2H_6), is mostly sourced from coal measures occurring at depths greater than those generally mined for coal (Willmott, 2017b). The most serious environmental issues associated with the CSG/LNG industry concern contamination of the region's groundwater and lowering of the water table with consequent effects upon other forms of land use particularly agricultural production. Such contamination may occur if the gas wells, charged with a mixture of gas and water, suffer leakage as this mixture passes through various strata to the surface where the gas is released. The functioning of the well and the efficiency of gas harvesting will vary with the nature of the geological formations through which the well passes and with the effectiveness of the well casing. The residual water after separation of the gas is contaminated and strongly saline, presenting problems of disposal since state regulations do not, at present, allow the use of evaporation ponds, whilst treatment by reverse osmosis would constitute a significant additional cost. Another complicating factor is that CSG harvesting, to be efficient, requires a number of extraction wells rather close together (Walker & Mallants, 2014).

The future of the industry will depend on the strict application of the state's water-use legislation and on the priority given to protection of Australia's most precious resource, the stored water of the Great Artesian Basin.

LEGISLATIVE ASPECTS OF MINE-SITE REHABILITATION

Mining (and its associated activities) makes a major contribution to Australia's economy, but it has also brought many environmental problems, many of which have proved to be long-lasting and costly to deal with. Managing these environmental impacts and the costs of rehabilitation of the affected areas is the responsibility of the appropriate state or territory governments through their legislative, land tenure and management regimes. In addition to the numerous currently operational mines of northern Australia, there is also a vast number of abandoned mine sites, many on state- or territory-owned land but many more on private land. The latter present a particular problem since there is often no existing tenure requiring the former miner to take responsibility for remediation or to address any site safety or health risks. For operational mines, the miner's legal responsibility for site rehabilitation are clear and should continue long after active resource extraction has been suspended.

There is, generally, a significant gap between the full remediation costs for a particular site and the total financial assurance (bond) held by the appropriate government authority (e.g. Queensland Audit Office data 2013–14, reported in Marlow, 2016). There are particular problems when the miner (or the body responsible for the environmental problem) goes into liquidation. In a recent development, the Queensland Government has amended its Environmental Protection Act, 1994 to allow the Queensland Department of Environment and Heritage Protection (QDEHP) to impose a chain of responsibility upon companies (and related parties) in financial difficulty so avoiding state (and hence taxpayer) liability for such costly clean-up bills.

An up-to-date example of the urgent need for this legislation was provided by the entry into voluntary administration of the Queensland Nickel–operated refinery at Townsville (see Chapter 1, Figure 1.4) in early 2016, leaving community concerns as to the continuing management of the reportedly toxic contents of the tailings dams and at the capacity of the company's remaining skeleton staff to handle the situation. The problem was further exacerbated by the fact that no financial assurance was held by QDEHP and by the proximity of the refinery to the environmentally sensitive Great Barrier Reef Marine Park and World Heritage Area. The amendment to the Environmental Protection Act, according to a report by the Environmental Defence Office Queensland (EDO Qld, 2016), ensures that persons "related" to a company (i.e. a person who has received financial benefit from the company's operations or has been in a position to influence the company's environmental conduct) can be issued with an environmental protection order (EPO) to ensure that funds are provided to undertake the necessary remedial action to avoid any environmental harm from the company's operations.

The current state of mine-site rehabilitation, specifically as it applies to Queensland mines, has been thoroughly reviewed by Marlow (2016), although it is likely that similar problems apply to all mine sites across northern Australia. Marlow reports that many aspects of the overall management and costing of the practices involved fall well short of what one might think would be community expectations. Interestingly, the guidelines created by the Western Australian Department of Mines and Petroleum and Environmental Protection Authority

(DMPEPA, Western Australia, 2015) for preparing mine closure plans were deemed to be the best currently available and were therefore recommended, with suitable revisions and modifications, for adoption more widely. Other aspects of mine-site rehabilitation that would significantly improve effectiveness of the processes involved would be the requirement that the granting of licences to operate a mine be conditional upon suitably approved site closure plans. The cost of rehabilitation, particularly in relation to assurance and royalty payments to the state/territory government, should also be fully transparent.

8 National Parks and Conservation

Australia has long had a policy of allocating areas of land for protection from development and their preservation close to what is presumed to have been their condition before the extensive land-use modifications brought by human occupation (Aboriginal and more recently non-Aboriginal). This is as true for the tropical north of the country as it is for the somewhat more densely populated southern regions. Indeed, the lower population density and the previously slower pace of development of the north have undoubtedly contributed to this more protective approach to land management in the north. In the state of Queensland alone, the most recent census listed over 300 areas designated as national parks, of which almost a half are located in the more sparsely populated tropical north. A similar recognition of areas designated as being of special environmental value has been evident in the Northern Territory and in Western Australia.

NATIONAL PARKS OF TROPICAL QUEENSLAND

Some of Queensland's national parks are located fairly close to population centres, while others occupy more remote (and in many cases environmentally unique) sites. In the Cape York Peninsula area, for example, no fewer than 26 national parks have been designated, many of them covering large and sometimes inaccessible areas. These and the vast array of other national parks, nature reserves, conservation areas and wilderness tracks represent an important resource and, as popular tourist destinations, make a significant contributor to the economy of the tropical north.

Acquisition and subsequent management of national parks are in the hands of the state governments, assisted by a range of other local government bodies and volunteer groups. The process of acquisition varies between different regions and depends on the previous ownership of the land and its particular former usage. In his publication *Five Million Hectares – A Conservation Memoir* Sattler (2014) gives a detailed account of the implementation process as it has been applied in Queensland since the 1970s. He describes the criteria upon which national park areas are selected and the conservation planning that informs their ongoing management.

In the formerly heavily grazed mulga (*Acacia aneura*) pastoral country west of the Great Dividing Range, for example, national parks have now become part of the landscape and accepted by the community as a legitimate land use. The acquisition process involves prior evaluation by the State Department of Lands, based on fair market value (as judged from sales of similar properties in the region) plus severance costs and other financial inducements or short-term transitional grazing rights.

Careful evaluation of the environmental significance of the parks, with an eye to outback tourism, has incidentally brought much-needed revenue to some hard-pressed rural communities. The establishment and subsequent destocking of parks across the major land systems is now being monitored to provide valuable information on rangeland recovery and reserve management.

Some national parks have been created by excision from much larger pastoral leases. The Boodjamulla National Park (~3,000 km²), established in 1985 in the gulf country of north-western Queensland, is situated in a region of large pastoral leases such as the Lawn Hill station, probably the largest in Australia. Just south of this station are the Riversleigh fossil fields, which have yielded valuable fossils from the region's rich Oligo-Miocene habitats and is now a declared World Heritage Site. Much of this area was originally pastoral land of inferior grazing potential, badly overgrazed and, in parts, suffering extensive gully erosion. The Moorinya National Park located 90 km south of Torrens Creek (20°53'S, 145°01'E) was acquired as a typical example of woodlands on Torrens Creek alluvial soils. After the purchase and subsequent destocking, it was discovered that an endangered small mammal species, the Julia Creek dunnart (*Sminthopsis douglasi*), had become re-established in the park (Mifsud, 1999; Woolley, 2008).

Many of the national parks of north-eastern Queensland are located in areas of former episodic volcanic activity, some of relatively recent origin (Willmott, 2009). Former lava flows to the west of Charters Towers and further north at Undara and in the Atherton Tableland area (see Chapter 1, Figure 1.4) now appear as partly collapsed lava tubes to form basalt walls. The Atherton Tableland, in particular, has many small national parks distributed within remnant patches of rainforest, some supporting crater lakes (such as Lakes Eacham and Barrine, both now included in the Wet Tropics World Heritage Area), originally created when rising basalt magma encounters groundwater in the surrounding rocks.

The establishment of national parks in tropical Australia has always proceeded in full accord with Aboriginal ownership. Over the period 2004–2014, over 2 million hectares in Queensland alone were placed under Aboriginal ownership. A large proportion of this is now jointly managed for conservation and traditional Aboriginal cultural activities. Under the Cape York Peninsula Land Use Strategy (CYPLUS), 1995, jointly funded by the Queensland and Australian governments, a number of agreements (later incorporating the Wild Rivers Act, 2005) were negotiated between Aboriginal, pastoral and conservation interests.

The creation of natural reserves in tropical areas previously devoted to sugarcane cropping might be expected to attract more opposition than would acquisitions from marginal grazing lands. Such acquisitions have, nevertheless, been achieved through what became known as the Sugar Coast Environmental Rescue Package (Sattler, 2014), which was much aided by public concern over threats to the survival of the endangered mahogany glider (*Petaurus gracillis*), which has a very limited range within the area of coastal north Queensland that had been scheduled for extensive expansion for sugarcane. The rescue package was jointly funded by the Queensland and Australian governments and allowed the appointment of environmental officers to liaise with the cane industry in regulating acquisitions and in developing ecologically sustainable production methods.

NATIONAL PARKS OF THE NORTH OF WESTERN AUSTRALIA

THE KIMBERLEYS

The Kimberley region of the north of Western Australia (see Chapter 1, Figure 1.1) can be defined as the area extending north from just south of latitude 18°S, taking in the widespread catchment of the Fitzroy River and continuing inland as far as the Northern Territory–Western Australian border (see Chapter 2, Figure 2.1). It is a region of spectacular scenery, much of it of particular historic and cultural significance to the original inhabitants. A number of sites have been granted national park status under the joint management of the State Department of Conservation and Land Management, the National Parks and Conservation Authority, and representatives of the traditional Aboriginal owners. Large national parks have been established at the Drysdale River area in the north and at the Purnululu area of south-east Kimberely, which includes the spectacular Bungle Bungle sandstone massif. Many nature reserves, conservation reserves and conservation parks, totalling an area of over 1.4×10^6 ha have also been dedicated (Western Australian Department of Conservation and Land Management, 1995).

The ongoing participation of the traditional owners in the management of the parks represents a formal recognition of the social, economic and cultural responsibilities that they bring to a better understanding of how the region should be developed and, where appropriate, conserved. Much of the area in the Ord River basin, for example, has been given over to a program of stabilisation and re-generation of vegetation following degradation by previous pastoral practices. Other areas have suffered from disturbance by feral animals and exotic plant invasions. Different management groups have different priorities for addressing land management issues leading to what seem to have become almost inevitable conflicts between the Aboriginal landholders and the state authorities or, as sometimes seems to be the case, between different Aboriginal groups. It is now widely recognised, however, that despite the developments of the last 100 years or so, the Aboriginal community of the region still maintains a strong cultural identity and attachment to the land and that they make a valuable contribution to the store of knowledge of the natural and cultural resources of the various national parks.

A state-sponsored survey of the wildlife and vegetation of the Pornululu (Bungle Bungle) National Park and adjacent areas (Woinarski, 1990) has provided a valuable basis against which the effects of more recent changes in land management practices, such as removal of stock and feral animals, can be assessed. Much of the rehabilitation of the area, especially in the Ord River catchment area is directed towards redressing the damage caused by the excessive grazing pressures inflicted over the early years of settlement by Europeans. This started early in the 1980s when the Western Australian government took control of those areas previously designated as pastoral leases. Mustering and removal of cattle were commenced and fencing erected to protect areas subject to strip contour cultivation of hardy pioneer species such as buffel grass (*Cenchrus ciliaris*) and birdwood grass (*C. setiger*). It is reported that Aboriginal people describe with sadness the changes to the land and the rivers since the arrival of Europeans in the Kimberley (Western Australian Department of

Conservation and Land Management, 1995). What were once large and abundant waterholes are now covered up and hold little or no water. Bush tucker, it is claimed, is increasingly difficult to find and some animals, once prized as important food sources, have become locally extinct.

THE PILBARA

The Pilbara region extends from Wyndham in the north to the Newman area in the south. There are several river catchments including those of the Fortescue and Ashburton Rivers which, like the other, mostly ephemeral rivers of the Pilbara, flow generally north-west to the Indian Ocean. The most extensive land use in the area is pastoralism, which occupies well over half of the total land area. A sizeable proportion of the remaining land, estimated by van Vreeswyk et al. (2004) to be well over 10% has, by now, been set aside for conservation. Many national parks, nature reserves and areas recently purchased by the Department of Conservation and Land Management (CALM) have been de-stocked and incorporated into the conservation estate. There are also many Aboriginal reserves and some special leases set aside for Aboriginal use as well as large tracts of unallocated Crown land.

The Millstream-Chichester Park located some 150 km south-east of Karratha (20°44'S, 116°52'E) has been described as an oasis in the desert since it has some permanent pools fed by springs that draw water from underground aquifers within the nearby porous dolomite rocks. The Millstream region is the traditional land of the Yinjibamdi people but was occupied by pastoralists from 1865 to 1967, when it became a national park with the original cattle property later adopted as the headquarters of the park rangers.

Large areas of the Pilbara feature widespread stony mantles on pediments, extensive nearly level plains subject to episodic sheet flow, tall shrub strata largely unaffected by grazing and widespread sandy plains with moderately dense spinifex grasslands (Van Vreeswyk et al., 2004). The floodplains and alluvial plains were judged to have been the most affected by past inappropriate land-use practices. The Pilbara flora is diverse with over 96% of the vascular plants native to the region. The most common species reported from the 2004 inventory were *Triodia pungens* (soft spinifex), *Acacia inaequilatera* (kanji bush), *Chrysopogon fallax* (ribbon grass), and the introduced *Cenchrus ciliaris* (buffel grass) (all perennials), while the most widespread annual species was *Aristida contorta* (windgrass). There had clearly been considerable degradation of perennial vegetation, mostly due to overgrazing. Only areas with a stony surface mantle or rock outcrops were reported to have escaped erosion.

One of the larger of the conservation areas is the Karijini National Park, located in the Hamersley Range just north of the Tropic of Capricorn. It is well known for its spectacular, rugged scenery and ancient geological formations and a wide range of arid land ecosystems. It is the traditional home of a number of Aboriginal clans and there is evidence of their early occupation dating back more than 20,000 years. Their management of the land, especially the fire-stick farming practices, resulted in the wide variety of vegetation types and stages of succession that characterise today's park. Further south in the Murchison and Gascoyne regions there are large areas of mulga (*Acacia aneura*) woodland, salt flats and boulder-strewn landscapes, including many degraded pastoral leases now acquired by the state government and thus adding substantially to the state's conservation estate.

NATIONAL PARKS OF THE NORTHERN TERRITORY

The Northern Territory has declared about 5×10^6 ha (c. 3.7% of its total area) as national parks or reserves. They include Kakadu (19,804 km^2), Litchfield (1,458 km^2) and Nitmiluk (2,924 km^2) parks (all situated in the far north) and a number of other smaller parks and reserves across the territory. Approximately 50% of the national park area can be classified as monsoonal savanna country and as such is subject to extensive fire outbreaks, either natural or controlled. The fire regimes of two of the territory's parks (Litchfield and Nitmiluk) have been monitored over an 8- to 9-year period using LANDSAT TM imagery supplemented, during cloudy periods, by appropriate alternative imagery (Edwards et al., 2001) and the interpretations checked against ground-truth data from the same area. It was found that over the study period, over 50% of Litchfield and over 40% of Nitmiluk was burned, the former burning substantially in the earlier, cooler and moister dry season, the latter mostly during the parched late dry season after August. Such frequency of burning, especially in the low open woodland/heath habitats was deemed to be ecologically unsustainable thus clearly requiring careful management, in terms of extent and seasonality (and hence intensity) of burning appropriate for the different landscapes across the north.

Remote sensing data from coarse-resolution NOAA-AVHRR imagery has provided information on the daily distribution of fires ("hot spots") over annual cycles across the northern savannas and has also allowed cumulative mapping of burnt areas ("fire scars") (Russell-Smith et al., 2003). It has been established that a great majority of annual burning occurs in the tropical savanna areas but with uneven distribution across a variety of major land uses (pastoral, conservation and indigenous). Most burning occurs in the latter half of the dry season, typically as uncontrolled wildfire.

JOINT MANAGEMENT OF NATIONAL PARKS

With the passing of the Aboriginal Land Rights Act (Northern Territory) in 1976, there ensued a clear need for a mechanism to arrange for a return to traditional ownership, land that had already been designated as national parks or conservation reserves. This was first accomplished for the Northern Territory through a system of joint management, that is, a form of legal partnership and management structure reflecting the rights, interests and obligations of the Aboriginal owners as well as those of the relevant government acting on behalf of the wider community. Similar but slightly different (in detail) arrangements were later established by the other states or, in some cases, by the Commonwealth Government.

Some of the major features of such joint arrangements have been described by Szabo and Smyth (2003). They include vesting of the land in trust on behalf of the traditional owners to be managed by a board of management whose members comprise a mix of traditional owners and government representatives. Arrangements may also be made for payment of an annual fee by the government to the traditional owners for use of the land as a national park. Day-to-day management of the park is usually the responsibility of the appropriate conservation or wildlife commission. The board and the commission, between them, prepare management plans, protect and enforce the rights of traditional owners, protect special sites and prepare management by-laws. In some cases, as appears to apply to the Kakadu National Park, there is provision for return of the land to Aboriginal ownership and simultaneously to a government conservation agency under the direction of the board of management.

9 Management Issues

Responses to the continuing changes impacting on the ecosystems of the tropical north of Australia will be largely in the hands of the landholders, guided by support facilities available within a governance framework at state, territory and national level, with essential input at the local community or landscape level. Van Oosterzee et al. (2014) have provided a comprehensive review of the range of land management initiatives as they have been applied in Australia in recent years, highlighting the changes that have occurred since the 1970s with the gradual change in emphasis from production-oriented systems to those addressing agricultural sustainability and natural resource management affecting water availability and quality, soils and pastures and biodiversity. They report on the increasing need for closer integration of multiple agro-ecosystem benefits at the regional level and decentralisation of management decisions and action to the lowest level of effective governance.

Such decentralisation would address some of the criticisms that have been made, for example, of the Australian Government's Caring for Country program, which was established in 2008 to build on the foundations established by the Natural Heritage Trust. Implementation of the Caring for Country program, it has been claimed (Robins & Kanowski, 2011), has not met the aspirations of regional organisations for core funding. Instead, priority seems to have been given to projects capable of demonstrating short-term, measurable outputs rather than to the support of local resource management bodies such as Landcare backed by the professional research input formerly provided by CSIRO Land and Water Australia.

A key factor in the sustainable development of the tropical north is the effective management of the region's ample, but seasonally and spatially variable freshwater resources. Many of the river catchments of the north have, in the past, been unregulated, and the water has been allocated on a licence-by-licence basis by state government agencies. In more recent times, water planning has become an essential part of any proposed development. The Northern Territory Government, a few years ago, announced a policy of capping water allocations to retain 80% of the natural flow in rivers and aquifers in the wet–dry region of its jurisdiction (Warfe et al., 2011). Such policies recognise that unimpeded flow regimes provide a range of ecosystem benefits and services to the region – such as the productivity of local commercial and recreational fisheries – beyond those directly related to consumption of the resource itself (Robins et al., 2005). The case for managing environmental flows in tropical rivers, as in rivers generally, rests on a conceptual understanding of the relationship between flow variation and ecological responses (Arthington et al., 2006).

The framework for flow allocations is defined as the ecological limits of hydrological alterations (ELOHA) (Naiman et al., 2002), which acknowledges the limits to which natural flows can be varied (e.g. by extraction) while still maintaining naturally functioning aquatic ecosystems. The floodplain rivers of northern Australia should, ideally, be managed in such a way as to sustain a range of waterhole types

from permanent and deep billabongs through to intermittent but annually inundated
stretches of creeks (Warfe et al., 2011). Management of river flows requires a reservoir of storage water that can be drawn down during (mostly annual) periods of water
limitation. For many areas of the tropical north, large water storages are, for a number of reasons, not really viable and it is likely that a more appropriate option would
be stormwater harvesting and managed aquifer re-charge (Clark et al., 2015) supported by small-scale developments including groundwater abstractions, off-stream
storages and tributary impoundments.

RURAL INDUSTRIES OF THE NORTH

Before 1999 most of the rural industries of the tropical north were represented by
three powerful bodies – the Cattlemen's Union of Australia, the Queensland Grain
Growers Association and the United Graziers Association – which merged to form
AgForce, an organisation with particular interest in the development, viability and
profitability of broadacre industries relating to cattle, grain, sheep and wool. Other,
more specialised rural industries, such as forestry and forest-products research, horticulture and cane farming, are represented by specific bodies largely supported by
government or industry funding.

AgForce was formed at a time of considerable change in the industry, a period of
some market turbulence and urban population drift as well as some uncertainty as to
the viability of the rural sector and its contribution to the economy and social fabric
of northern Australia. In collaboration with state and commonwealth governments,
AgForce now concerns itself with issues of resource management, land tenure, environmental impacts, international competitiveness and rural community services. It
is involved in formulating (and amending) vegetation management laws based on
negotiation between landholders and state government. It takes part in assessing
applications for high-value irrigation projects and in any appeal processes arising
from such applications. It strongly supports extension workshops for landholders to
refresh knowledge and understanding of their rights and responsibilities.

Among the challenges faced by those charged with stewardship of the north's rural
industries are the undeniable fact of the recurring and probably future increasing
incidence of natural disasters, calling for government-funded emergency responses.
Such responses are most effective when the emphasis is on prevention and mitigation
rather than recovery and rectification. Drought conditions, for example, affect the
yield of dryland crops and make heavy demands on stored water. In years of exceptionally low rainfall many crops are having to be turned over to grazing without ever
reaching the harvesting stage. Other management decisions calling for urgent attention and government involvement include the development of biosecurity protocols
to combat serious incursions such as foot-and-mouth disease, equine influenza, red
fire ant and other pest infestations.

The complexity of the changing demands now placed upon the rural industries of
the north have been reported by Turner and Lambert (2016) specifically in relation
to forestry and forest-products research. They describe the change in management
practices that occurred in 1985. Before that date, these activities were largely under
the stewardship of state governments, with additional research support provided by

the Commonwealth Government through CSIRO and the universities. There were also strong links with timber producers and timber processors. After that date, they note, the linkages between forest management and research organisations were greatly diminished with accompanying (and perhaps because of) reduced expenditure and loss of scientific, technical and support staff.

INDIGENOUS AUSTRALIANS AND TROPICAL LAND MANAGEMENT

A common view during the early part of the 20th century was that Australia's Aboriginal nations during their sole occupancy of the land had moulded their lifestyle to their environment, reflecting the intimate relationship between them and their homeland. They were considered to be a society living in harmony with nature. From the 1970s onward, however, there is a growing body of anthropological evidence that the pre-1788 Australians were skilled managers of the land and, in particular, were expert in the use of fire as a tool to control the vegetation and indeed the entire ecosystem. An alternative view that the Aborigines observed and made use of an existing natural fire regime rather than attempting to develop a new one has also been put forward (e.g. Horton, 1982).

Gammage (2011) has presented a wealth of valuable information describing how Aboriginal people "shepherded fire around their country – caging, invigorating, smoothing the immense complexity of the country's plants and animals". In this way they were able to prevent the uncontrolled wild fires that, especially after several good growing seasons and the build-up of flammable material, would otherwise sweep over large areas destroying everything in their path.

Observations by early non-Aboriginal explorers of the north and the evidence of oral history passed on by the elders confirm that one of the major aims of Aboriginal land management was the control of fire. In east Arnhem Land (see Chapter 1, Figure 1.1), for example, it was noted that as soon as the grass crop begins to dry, controlled burning was started usually in conjunction with organised hunting drives. Firing took place only in limited areas and always under strict control, with particular care to protect the vines of food plants. Over vast areas of the north, a succession of fires would create a mosaic of recovery stages, controlling the spinifex without killing the trees and protecting other wildlife. It soon became apparent that the open forests reported by the early explorers were not a natural association but a secondary one, the result of the careful land management skills of the original inhabitants. Some of the early illustrations show clear lines of demarcation between open forest and vine scrub as were noted, for example, by Captain Cook during his enforced extended stay in the Endeavour River region.

Certain areas of coastal Cape York Peninsula (see Chapter 1, Figure 1.1) have been described by other early visitors as possibly some of the richest and most varied environments for hunter-gatherers in the world (Chase, 2009), supporting a form of land management that included hunting, often with the aid of dogs and nets, for snakes, crocodiles, lizards, echidnas, parrots, cockatoos, wildfowl, pigeons, cassowaries and emus. Firestick farming methods were used to drive animals into ambushes and to encourage grazing on grasslands created by controlled burning. Such clearings (balds) have been reported across Cape York and elsewhere in the north while some similar areas, formerly providing yams and root, have now reverted to rainforest. Aboriginal fishing grounds have been described as extending from coastal areas far out into the Barrier Reef lagoon, using outrigger canoes to harvest dugong, turtles, rays and a variety of fish.

The present-day challenge for land management of the tropical north is to devise a form of rural development that can deliver the level of productivity demanded by its current (and predicted increased) population without compromising the degree of sustainability that appears to have existed under the stewardship of the much more sparsely populated pre-1788 Aboriginal occupants. The extent to which this is possible or even desirable is a contentious issue. Some have argued, for example, that fire management practices as used by Aboriginal land managers are no longer appropriate for today's tropical north. It has to be recognised, as Whitehead et al. (2003) have pointed out, that the objectives of the Aboriginal land managers and the values they applied to savanna landscapes would have been quite different from those of present-day land managers, whether they be Indigenous or non-Indigenous. The aim today is more likely to entail some form of application of Aboriginal prescriptions for tropical landscapes to suit the changed land management objectives of today.

WILD RIVERS ACT

The Wild Rivers Act, declared by the Queensland Government in 2005, had as its main purpose the preservation of the natural values of certain rivers, which, at the time, had most of their natural status still intact. It also sought to give local communities greater involvement in management decisions to ensure protection of areas of high conservation value. Under the act, certain river basins – most of them on the Cape York Peninsula but including rivers in the Lake Eyre basin and the Channel Country – would be protected from large-scale development. The act passed through the state parliament with bipartisan support but was later subject to strong criticism, especially from those claiming that Indigenous people and the traditional owners had not been consulted prior to its passage into law (Neale, 2017). As such, it was said to violate both native title principles, as specified under the 1996 Wik agreement (Brennan 1998) and the principles of consent outlined in the United Nations Declaration on the Rights of Indigenous People.

The Carpentaria Land Council (CLC) challenged the act in 2009 and the Federal Court, in 2014, ruled in its favour. Before its repeal, however, the act was the basis upon which the Cape Alumina's proposed Pisolite bauxite mine near Mapoon on the north-west coast was blocked because a significant amount of the bauxite was found to be located within the Wenlock River's buffer zone (Neale, 2017). The act was later replaced by a regional plan that retained many of the Wild Rivers' provisions but with significantly reduced recognition of Indigenous rights.

BUFFALO AND WILD PIG IN ARNHEM LAND

In the 1980s the federal government established its Brucellosis and Tuberculosis Eradication Campaign (BTEC) across large areas of northern Australia. This called for the eventual elimination of wild buffalo and feral cattle regarded, at the time, as a threat to the beef export trade (Altman, 1982). For a variety of reasons, including the significance of buffalo as a source of meat to the community, the BTEC did not extend into Arnhem Land. Within a decade, however, and following the formal establishment of a community-based conservation group, the Djelk Rangers, serious concerns were raised at the visible damage from the rapidly growing population of wild pig on flood plains, billabongs and riparian margins. These concerns were later extended to include the wetlands-choking exotic plant *Mimosa pigra* and by the late 1990s to what had by then become the ever-growing herds of buffalo. By this time, the national government had established its Indigenous Protected Areas (IPA) program and a number of collaborative research and monitoring programs had been launched, including the Australian Research Council Key Centre for Tropical Wildlife Management and the Charles Darwin University-based Cooperative Research Centre for Tropical Savannas Management (Altman, 2017). An aerial survey carried out in 2014 to determine the distribution and abundance of buffalo (and other large feral vertebrates) in Arnhem Land estimated that buffalo numbers in that region had reached numbers close to 100,000 but with large variations across different ecological zones. As a consequence of the declaration of the Kuninjku IPA (covering over 1,000 km^2) the clans concerned were required to deliver various environmental "outcomes", especially given that many of their numbers were employed as salaried rangers under the federally funded Working for Country program or engaged as "custodial consultants" assisting rangers. In particular, they were faced by what by then had become the evident environmental and biodiversity threats posed by the increasing buffalo and pig population explosion (Altman, 2017).

At the same time, the Kuninjku clans had become increasingly dependent on buffalo (and to a lesser extent pig) as a source of meat due, no doubt, to the decline in the availability of other bush foods. Plans to control buffalo numbers by environmental culling or managed commercial harvesting came up against perceptions of buffalo as a valuable natural resource rather than a feral species (like wild pig, cane toad and the invasive weed *Mimosa pigra*) and as a species that for over 200 years has become adapted to the conditions of the perched wetlands of east Arnhem Land. Others have argued that even if buffalo is not endemic, Kuninjku clans may have special native title property rights relating to this species. Thus, according to the Native Title Act, for example, landowners are granted the right to harvest species for domestic or non-commercial use. Such considerations highlight the difficult challenges of engaging with the rights of the indigenous community within a conservationist framework. Langton (2012) has reminded us that Aboriginal people, in addition to their relationship with the land also have an economic life and, like everyone else, are caught up in the transforming encounter with modernity.

DOMICULTURE PRACTICES OF CAPE YORK PENINSULA

Domiculture is defined as the framework within which Aboriginal groups sharing cultural beliefs and practices operate within their localised physical and biotic environment (Rindos, 1980). The term has been applied by Hynes and Chase (1982) to describe the land-use practices of Aboriginal groups of the east coast of Cape York Peninsula. The area in question (north of the 14°S latitude) consists of a coastal plain varying in width from 5 to 20 km to the west of which lie the hills and mountains of the Great Dividing Range. To the east, numerous islands, cays and spur reefs are scattered across the sheltered waters of the Great Barrier Reef, which, in this region, is scarcely more than 20 km offshore.

The coastal plain consists of a mosaic of open and closed woodlands, heaths, riverine vine forests, swamps, open grasslands and littoral scrubs. During the monsoon season (January to May) the coastal swales, swamps and waterways are flooded and vegetation growth is vigorous. During the dry season, surface water diminishes and by December only a few permanent swamps and watercourses remain. Both land and sea are rich in food resources. The seasonal cycle brings large migrations of birds and fishes, and the plants yield continuous and overlapping fruit throughout the year. The area was once occupied by a number of Aboriginal groups, some descendants of whom now reside at the Lockhart River Aboriginal community. The Aboriginal communities retain an extensive knowledge of the plants and animals of the region as well as a detailed knowledge of their responses to seasonal changes. This would include the practice of "farming" and harvesting yams (*Dioscorea sativa* var. *elongata*) as has been described to occur in some other tropical regions (Jones, 1975). Coconuts may also be collected from beaches and planted above the tidemark as part of a more widely based planting and management program that includes a wide range of species.

As a result of their analysis, Hynes and Chase (1982) concluded that the hunter-gatherers of the Cape York area, like those from other regions of the north, were not ecologically passive and were not moulded by a largely "natural" landscape. Neither can it be assumed that the hunter-gatherer modifications of the landscape are the result of "accidental" or unconscious action. Rather, the domicultural plant communities are the result of Aboriginal management, which, although no longer actively practised, may yet hold lessons for current agricultural processes in being less vulnerable to environmental perturbations and perhaps, disease.

10 Conclusions

The ecosystems of the tropical north of Australia are supported by a rich flora with many unique features reflecting their geological and climatic history. They occupy part of a region that has long been isolated from other land masses and, until recent times, has been sparsely populated compared with most other habitable regions of the world. Of all the more recent changes experienced by the ecosystems of the tropical north, two stand out in terms of the scale of their impact and of the rate of change. The first is that wrought by the unprecedented development associated with the "opening up" of the north that followed on from the colonial and post-federation developments of sub-tropical and temperate Australia. The second is that due to global climate change, also linked, it is now generally accepted, to post-industrialisation influences. Both changes, in the absence of effective mitigating measures, are likely to continue into the future at rates exceeding any that have occurred over the most recent 5,000-year post-glacial period.

It has been predicted that with average global atmospheric CO_2 concentrations committed to continuing increases, consequent increased atmospheric temperatures are on a path leading to a warming of close to 2°C above pre-industrial levels (Steffen et al., 2009) within decades, unless deep cuts in global greenhouse gas emissions can be achieved. Projections for Australia (for 2030 and 2050) based on time series changes of temperature and rainfall indicate a general warming, with increasing rainfall over northern, central and north-west regions and decreasing rainfall in the eastern, south-eastern and south-western regions (Barron et al., 2011). Some of the predicted changes are already apparent as a southerly extension of the tropical climate zone of the far north and a contraction of the northern extent of the arid zone. Expansion of the arid zone to the south and south-east would indicate a possible contraction of the northern extent of the southern cropping zone and a gradual change of land cover from annual cropping to native shrubland and grassland. The impact of the predicted climate changes on the tropical zone is, according to Barron et al. (2011), more difficult to predict because of the greater uncertainty due, in particular, to regional variations of rainfall.

On a global basis, it is clear that offsetting the ever-increasing atmospheric CO_2 concentrations, and thereby providing a bridge between the present and the planned-for low-fossil-fuel future, can only be achieved through the process of photosynthetic carbon fixation. This focuses on those ecosystems with the greatest capacity for photosynthesis (and carbon storage) such as tropical forests, woodlands and savanna lands, wetlands and offshore marine ecosystems. Tropical forests, for example, estimated to cover no more than 12% of the earth's land surface, are nevertheless responsible for well over 30% of gross terrestrial primary production (Bonan, 2008), pointing to their immense importance to the global carbon cycle. However, to make a significant contribution to reducing atmospheric CO_2 concentrations, the photosynthetically fixed carbon has to be locked away as storage carbon rather than be subject to decomposition or further metabolism.

Photosynthetically fixed carbon that is allocated to fine roots and leaves has a high turnover, decomposes rapidly and thus returns quickly to the atmosphere. Similarly, crops and pastures, and other plants grown for consumption make only a limited contribution to carbon mitigation. Carbon allocated to wood, on the other hand, is fixed for decades or even centuries (Luyssaert et al., 2008). The pattern of fractional allocation of the products of photosynthesis is thus a major driver of long-term carbon sequestration, creating the carbon pool represented, for example, by tropical forests (Carbone et al., 2013), especially in the larger trees (Slik et al., 2013).

There has been considerable discussion of the possible mechanisms governing the allocation of photosynthetic biomass within trees. The pipe model (Shinozaki et al., 1964) assumes a more or less constant relationship between sapwood and leaf area, thus maintaining a balance between transpiration and water transport. Superimposed upon such a functional balance would be an allocation to carbohydrate reserves to provide for drought, herbivory, diseases and disturbances and to support re-growth after periods of stress. An alternative hypothesis interprets any allocation to reserves simply as the difference between productivity and organ (or other) demand, with no active regulation. This implies a hierarchical method of allocation, which assumes that leaves, fine roots and reserves have a higher priority than sapwood and that assimilates are allocated to sapwood only after leaves, fine roots and reserves have reached their optimal size (Schippers et al., 2015).

Houghton et al. (2015) have assessed some of the factors governing the capture and storage of carbon by tropical forest ecosystems. They have estimated that gross emissions from forest can be up to two or three times greater than net emissions, suggesting that there is scope for improved management techniques either to enhance carbon uptake or to reduce emissions. It has been estimated that achievable changes in forest management could account for as much as 50% of total carbon emissions and, moreover, could be implemented more quickly than any industrial transition from fossil to renewable fuels. Among the management decisions most likely to contribute significantly to reducing carbon emissions would be reafforestation of low-yielding agricultural land and avoidance of deforestation and land clearing for "swidden" (i.e. slash-and-burn) agriculture.

The rate of carbon accumulation in forests diminishes as they mature in a process which, conservatively, could last for 50 or so years before declining to zero. Many tropical forests are currently growing (and accumulating carbon) and will continue to do so for some decades as long as they are subject to astute management (including selective logging) and improved agricultural practices. General support (e.g. from United Nations Climate Change Conventions) for Reduced Emissions from Deforestation and Forest Degradation (REDD) schemes worldwide indicate an acceptance of the central role of tropical forest management in reducing atmospheric CO_2 concentrations. Above-ground forest biomass constitutes a pool of carbon that is, however, vulnerable to droughts, fires, insects and other disturbances and, at best, can provide no more than a temporary carbon sink that can be effective only alongside an orderly transition away from fossil fuels.

Next to tropical forests, the other major potential carbon sink is that represented by the offshore and oceanic waters of the region and, more widely, by the Southern Ocean generally. Some indication of the potential for carbon sequestration in such waters can be obtained from the results of iron-enrichment experiments which have produced large phytoplankton blooms and hence an increased capacity for carbon export and possible sequestration (Boyd et al., 2007). Natural inputs of iron and major nutrients (e.g. from upwelling currents) have shown even more dramatic effects (Blain et al., 2007) indicating a considerably increased potential for removal of CO_2 from the atmosphere. The extent to which this form of fertilisation represents successful sequestration/mitigation is more questionable.

To be deemed successful as a form of mitigation, the sequestered anthropogenic carbon must remain in the deep ocean (and isolated from the atmosphere) for an extended period. The Intergovernmental Panel on Climate Change (IPCC) has recommended a standard 100-year time horizon and Robinson et al. (2014) have conducted simulations at oceanic depth over this time period. Their calculations indicated that 66% of carbon from a depth of 1,000 m returns into contact with the atmosphere within 100 years. That originally sequestered at 2,000 m depth, however, experiences a lower (29%) leakage back to the atmosphere. Moreover, carbon sequestered in the Southern Ocean becomes redistributed throughout the world's oceans within 100 years, largely due to the chaotic nature of the Antarctic Circumpolar Current flow. Clearly, physical transport is as important as the biogeochemical processes to the efficiency of oceanic carbon sequestration.

The fragility of coastal ecosystems such as coral reefs, mangroves, sea grasses and algal or plant primary producers to climate and anthropogenic changes puts in considerable doubt their capacity to make an effective and continuing contribution to global carbon sequestration.

Australia's tropical ecosystems, in facing the combined stressors of an increasing population and land-use development as well as the continuing effects of global climate change, will undergo changes calling for strong and, in many cases, novel forms of land management. There will be particular threats to their species richness and to their role as contributors of genetic resources. Changes due to land clearing, cropping, grazing and the introduction of exotic and invasive plant species, already evident in many ecosystems, will threaten their sustainability and the resources upon which they depend. There will be further threats from increased use of fertilisers, changing fire regimes, as well as greater urbanisation, mining and tourism.

Northern Australia shares with other tropical regions of the world a potential for high levels of agricultural productivity but with severe constraints of resource availability that can only be met by imaginative land use, management and investment. It will face strong pressures for continuing expansion of agriculture to meet demands for food, fibre and biofuels, but expanding the agricultural footprint of northern Australia will have to take into consideration the unique ecology of the region and its vulnerability to inappropriate development. We already have numerous examples of overexploitation of the available resources such as overstocking of grazing lands, inappropriate timber harvesting and deforestation, overfishing, poor conservation of natural water resources and contamination (and in some cases extensive depletion) of groundwater reserves. In too many instances in the past, development and exploitation of the region has been dominated by a mining mind-set that gives priority to harvesting the product in the most economical way with scant regard to sustainability and the preservation of natural values.

Such an approach to management of the natural and developed northern ecosystems is reflected in the well-documented decline of species diversity and in the number of reported species extinctions, with many more feared (Steffen et al., 2009; Woinarski et al., 2011). The ability of ecosystems to respond to rapid changes is clearly limited. The most that can be expected is that through a program of regional adaptive management, some form of self-adaptation can be established while maintaining well- functioning ecosystems (some with novel composition) that still deliver ecosystem services and, where appropriate, agricultural productivity. Such programs, primarily the responsibility of the states and territories, are now increasingly managed through commonwealth–state processes but devolved to national resource management (NRM) programs of catchment authorities and land-care groups. They will require new partnerships involving land managers, commercial interests, environmental bodies and Indigenous landowners.

Tropical Australia has witnessed most of the changes encountered by other parts of the tropical world (increased migration, trade liberalisation, greater urbanisation) but compressed into a shortened time frame. In the context of food production and human health and well-being, the original indigenous occupants of the region have had to adapt over a few generations from a hunter-gatherer lifestyle to a very different form of agricultural production and land management. Over this time there have been major shifts in the availability, affordability and acceptability of different types of food. This is seen most clearly in the more remote areas, especially in those areas populated by Indigenous communities, but other communities have faced similar problems. Indigenous and non-Indigenous communities alike suffer from a number of nutritional inadequacies ranging from malnutrition to dietary and health problems.

Such conditions, as have been described by Gillespie and van den Bold (2017) for world agriculture generally, are due to dysfunctional interactions between the agrifood systems, environmental factors, the health services and, crucially, the system of individual and household decision-making. Agriculture is clearly the key component in all this, but it interacts with a number of other activities such as social protection to alleviate the effects of seasonality, climate shocks and various other stresses. Agriculture provides food and is also a primary source of employment and income for rural communities. While agriculture offers relatively high returns on investment, there remains a disconnect between agricultural production (crops and livestock) and improved health and well-being of the community.

Globalisation has generated marketing systems that require intensive, more standardised and more costly production systems with longer and more complex supply chains. Local and family-based agriculture and its associated agro-biodiversity are being marginalised with end results often reflected in major and rapid shifts in dietary patterns. All of these trends increasingly threaten (and contribute to) environmental changes such as those due to global warming, desertification and the use of food crops for non-food purposes. Greater industrialisation of agriculture and a move towards a more restricted range of high-yielding varieties with accompanying loss of food diversity together with increasing use of agricultural chemicals as nutrients or for plant protection bring risks of groundwater depletion and soil and water contamination. Climate change effects, whether or not attributable to industrial and/or agricultural emissions, are all likely to add significantly to agricultural costs, especially when compounded by weather-related shocks or harvest failures.

The contribution of tropical agriculture to Australia's growing and diversifying production of food and food products for the region is predicted to increase. Beef and lamb exports to Asia from Australia generally are already spreading to northern areas and will continue to meet growing demands from increasing and more profitable south-east Asian markets such as Singapore. The live cattle industry, despite uncertainties of the Indonesian trade, largely driven by the desire of the Indonesian government for greater self-sufficiency of food supply (Trewin, 2014), is likely to continue, at least for the foreseeable future. Collaborative research programs funded by the Australian Centre for International Agricultural Research (ACIAR) have indicated high-value returns from measures aimed at improving the beef supply chain. Demand for beef in other south-east Asian countries such as China and Vietnam, is growing rapidly and, for economic reasons, is likely to continue to be preferred over processed beef by some importers.

Another predicted response to climate change is a decrease in diffusive groundwater recharge over most of western, central and south-eastern Australia but an increase in recharge across the north and parts the eastern region. The latter would be strongly influenced by the hydro-geological settings of the aquifers and by the level of groundwater use, which in most areas is mainly for agriculture, followed by domestic and town water use and finally commercial and mining use, although in the Pilbara region of Western Australia, the major groundwater demand comes from the mining industry.

Mine-site rehabilitation is an ongoing and growing problem for northern Australia. The costs involved invariably far exceed any budget set aside for restoration and rehabilitation of the ecosystem or even simply to minimise risks to the environment or to the health and safety of the community. There is now an increasing acceptance that the chief priority should be the successful establishment of a vegetation cover that can maintain itself and sustain a viable ecosystem rather attempting to replicate an ecosystem equivalent to that removed by the original clearing of the mine site.

Studies by Gould (2010) at the Weipa bauxite plateau on the Cape York Peninsula have compared the composition, abundance and richness of bird assemblages of rehabilitated sites (at various stages since mining started in the 1960s) with that of equivalent native forest sites. It was clear that mining and post-mining rehabilitation have resulted in habitat conversion rather than restoration. Indeed, there was evidence that the rehabilitation process had created new habitats (in this case for new, more generalist, bird species rather than the original habitat specialists). Even after more than 23 years of rehabilitation it was evident that some native forest bird species had still not returned, an indication, perhaps, that rehabilitation, however well conducted, may never entirely be able to offset biodiversity loss.

One of the imperatives for a sustainable healthy environment is the maintenance of biological diversity, which, in the tropical north of Australia, inevitably requires the active participation of one of the major landholder groups, the Aboriginal community. In the Northern Territory, for example, Aboriginal people have exclusive title to almost a half of the land and to long stretches of the coastal regions as well as non-exclusive native title rights over much of the remaining land.

The participation of representatives of various Aboriginal groups in planning different forms of land development has already had a moderating effect on both the pace and nature of such developments. This is likely to continue into the future as also are various economic considerations such as the availability (and source) of investment funding. The Northern Territory government as recently as last year announced that the proposed expansion of the Ord River Irrigation Area (ORIA) irrigation project has been put on hold. Stage 3, which had earmarked 14,500 ha of land near the Western Australia border for development, will now be delayed and perhaps modified, notwithstanding that it had earlier been included in the federal government's white paper for development of northern Australia and had been the subject of a feasibility study. Negotiations involving the Northern Territory government, native titleholders and the Northern Land Council raised concerns over the particular insistence of the Land Council to maintain the Keep River Plain as a sacred site.

The second stage of ORIA already under way and involving the Chinese company Kimberley Agricultural Investment (KAI) is earmarked for further expansion. It has been suggested that development of stage 3, if it proceeds, would provide economies of scale that would allow expansion, perhaps to include sugar cane planting. KAI and existing growers in the valley already have a viable grain industry with good prospects for the future. In the meantime, the Northern Territory government has reaffirmed its commitment to development of the Katherine region following trial plantings of soybeans, asparagus and other crops. But its commitment to support technical planning for ORIA stage 3 continues with the purchase by KAI of more land for development on the Western Australian side of the border.

Fire, either from anthropogenic or natural sources, will always be a factor in land management in the north (Tothill, 1971). Its use will continue to be crucial to the viability of the grazing industry and to the success of conservation and management of national parks and reserve areas. The opportunities for Indigenous people to apply their traditional land-use practices to more recent land management in the tropics have been limited. After the introduction of cattle in the 1850s and until fairly recently, Aboriginal people worked alongside non-Indigenous cattle station workers as stockmen and may also have used fire-related measures in stock management (Yibarbuk et al., 2001). Studies at the central Arnhem Land region of the Northern Territory have confirmed that maintenance of the ecological integrity of the plateau owes much to the continued application of traditional fire management practices which have suppressed the growth of annual grasses (*Sorghum* sp.) and limited the accumulation of fuel in the form of perennial grasses (*Triodia* sp.) and vegetation litter. Clearly, the skilled form of fire-based land management, as practised by the original occupants of the land may still have relevance to the lives of modern Indigenous communities despite the rapid and ongoing transformation of their traditional lifestyle.

While Indigenous Australians have been employed in the forestry sector since early European colonisation, there is clearly scope for greater participation in a range of forest-related activities from conservation to production (Feary et al., 2010).

In the emergency response program initiated under the National Indigenous Forestry strategy following the damage caused by Cyclone Larry in northern Queensland (2006), Aboriginal participation in the related Operation Farm Clear was deemed to have been very successful (Loxton et al., 2012).

Many of the contemporary land management practices in relation to the use of fire show a clear link with traditional Aboriginal methods and it is likely that future land management, especially in the context of reducing CO_2 emissions from biomass burning (Russell-Smith et al., 2009), will find much relevance in Aboriginal experience. In particular, input from Aboriginal fire-management techniques has the potential to influence calculations of greenhouse gas (GHG) emissions from savanna burning and may even help to moderate such emissions. It has been estimated that despite the 1.9×10^6 km^2 total area of northern Australia's tropical savanna region, it is responsible (for inventory purposes) for no more than between 1 and 3% of the country's total GHG emissions.

There is a clear need for an integrated form of land management, especially in managing the complex of responses that will be required to deal with the challenges of continuing climate change impacts. The global nature of the problem implies that the planning and costing cannot be left to the rural and regional communities but must involve integration of multiple service benefits under centralised regulation and funding. This would build on existing land-care groups and extension of the concept of local stewardship involving cooperation between the conservation and productive or agriculture sectors in formulating and implementing appropriate management plans. State and national government involvement such as that already established through various regional resource management bodies will be vital, especially given the complexities and demands of any carbon-pricing mechanisms that is implicit in any land management decisions.

POSTSCRIPT

In early February 2019, after this chapter had been completed, a vast region of the tropical north of Queensland experienced a torrential rainstorm resulting in extensive flooding of the rivers flowing east into the Coral Sea (causing considerable flood damage in the City of Townsville, population ~180,000) and extending along a 700 km stretch of the coast (from Cairns in the north to Mackay in the south) and inland to link up with the north-west flowing rivers entering the Gulf of Carpentaria.

The extent of the flooding and, in particular, its impact on the northern grazing industry prompted comparisons with previous floods in the region, such as that of 1974. In that year, when the inundation created an inland sea approximately 200 km across, the floods were described as "the worst in whiteman's memory" and stock losses were estimated as over one million heads. The floods of 2019 were of shorter duration than those of 1974 but were described as being more intense with many record daily and weekly rainfall totals. Stock losses in 2019 were also more localised. In the Normanton (17°40'S, 141°04'E)/Karumba (17°29'S, 140°04'E, north of Mount Isa; see Chapter 1, Figure 1.1) and in the Flinders River area of the gulf country, an estimated 75% of the cattle were lost to flooding, most dying of bogging and pneumonia (Source: Queensland Country Life). Losses from individual paddocks could be as high as 90% (e.g. 338 dead out of a total of 350 at Channel Downs, near Julia Creek, 250 km west of Hughenden [see Chapter 1, Figure 1.4], where rainfall was recorded as 735 mm over 10 days).

Interpreting the differences between these two flood events as an indication of a global climate trends is very difficult. For one thing, the years leading up to the 1974 floods were described as good grass years (i.e. strong cattle in good condition), whereas the years preceding the 2019 floods were record-breaking drought years. Moreover, the capacity of the present-day local population (and the available emergency services, including military and helicopter assistance) to respond to the crisis would undoubtedly have improved, although completion, in 1983, of the second stage of the Ross River Dam (with flood mitigation as one of its stated prime functions) appears to have been of only marginal impact since, for safety reasons, substantial releases over the spillway had to be made at the height of the flood.

Flood recovery programs, funded by the federal government, gave initial priority to safe disposal of cattle carcases and repair of damaged fencing but later addressed recovery packages to assist with re-stocking, aided by government grants or low-interest loans, the latter in collaboration with the banking sector. Other options under consideration included scaling down of the stock holdings and in extreme cases, the provision of assistance to allow the landholder to move on, with consequent implications for social issues such as rural population decline and loss of amenities for the local community.

Such flood recovery costs, when considered alongside the costs of prolonged drought, crop storm damage and disease outbreaks (all directly or indirectly linked to extreme weather conditions), have revived discussion of the concept of stewardship payments to landowners, defined as reimbursements for their role as custodians of the land. There is no doubt that there is considerable public sympathy for the problems faced by rural landholders and a growing understanding of the cost of maintaining healthy ecosystem services. Stewardship payments, perhaps via regional community catchment bodies (Edwards, 2014), would, it was argued, provide landholders with a regular income stream to cushion against unexpected climate-related costs and to assist in the management of risk and insurance arrangements.

References

ACIAR. (2012a). Enhancing smallholder benefits from reducing emissions from deforestation and forest degradation in Indonesia. ACIAR Project FST 2012/040.

ACIAR. (2012b). Teak-based agroforestry to enhance and diversify smallholder livelihoods in Luang Prabang province, Lao PDR. ACIAR Project FST 2012/041.

ACIAR. (2012c). Biological control of galling insect pests of eucalyptus plantations in the Mekong region. ACIAR Project FST 2012/091.

ACIAR. (2014). Maximising productivity of Eucalypt and Acacia plantations for growers in Indonesia and Vietnam. ACIAR Project FST 2014/064.

Adams, V. M. & Moon, K. (2013). Security and equity of conservation covenants: Contradictions of private protected area policies in Australia. *Land Use Policy* 30:114–119.

Aldrick, S. J., Buddenhagen, I. W. & Reddy, A. P. K. (1973). The occurrence of bacterial leaf blight in wild and cultivated rice in northern Australia. *Australian Journal of Agricultural Research* 24:219–227.

Alexander, D. McE. & Possingham, J. V. (1984). Potential for fruit growing in tropical Australia. In: *Tropical Tree Fruits for Australia*, pp. 1–9. Queensland Department of Primary Industries, Brisbane, Queensland, Australia.

Allen, C. D., Macalady, A. K., Chenchouni, H. et al. (2010). A global overview of drought and heat-induced tree mortality reveals emerging climate change risks for forests. *Forest Ecology and Management* 259:660–684.

Allen, C. G. (1986). Rum Jungle, a perspective. In: Van Groenou, P. J. R. & Burton, J. R. (Eds.), *Environmental Planning and Management for Mining and Energy*. Northern Australian Mine Rehabilitation Workshop, Darwin, Northern Territory, Australia, 7–12 June 1986. Darwin: Department of Mines and Energy.

Alongi, D. M. (2014). Carbon cycling and storage in mangrove forests. *Annual Review of Marine Science* 6:195–219.

Altman, J. C. (1982). Hunting Buffalo in North-Central Arnhem Land: A case of rapid adaptation among Aborigines. *Oceania* 52:274–285.

Altman, J. C. (2017). Kuninku people, buffalo and conservation in Arnhem Land: "It's a contradiction that frustrates us". In: Vincent, E. & Neale, T. (Eds.), *Unstable Relations: Indigenous People and Environmentalism in Contemporary Australia*, pp. 54–91. UWA Publishing, University of Western Australia.

Anderegg, E. R. L., Klein, T., Bartlett, M. et al. (2016). Meta-analysis reveals that hydraulic traits explain cross-species patterns of drought-induced tree mortality across the globe. *PNAS* 113:5024–5029.

Antal, M. J. & Grønli, M. (2003). The art, science and technology of charcoal production. *Journal of the American Chemical Society* 42:1619–1640.

Anthony, K. R. N., Maynard, J. A., Diaz-Pulido, G., Mumby, P. A., Cas, L. & Hoegh-Guldberg, O. (2011). Ocean acidification and warming will lower coral reef resilience. *Global Change Biology* 17:1789–1808.

Apgaua, D. M. G., Ishida, F. Y., Ting, D. Y. P. et al. (2015). Functional traits and water transport strategies in lowland tropical rainforest trees. doi:10.1371/journal.pone.0130799.

Arthington, A. H., Bunn, S. E., Poff, N. L. & Naiman, R. J. (2006). The challenge of providing environmental flow rules to sustain river ecosystems. *Ecological Applications* 16:1311–1318.

Ashton, P. S. (2003). Floristic zonation of tree communities on wet tropical mountains revisited. *Perspectives in Plant Ecology, Evolution and Systematics* 6:87–104.

Atienza, S. G. & Rubiales, D. R. (2017). Legumes in sustainable agriculture. *Crop & Pasture Science* 68:i–ii.

Australian Government. (2011). Carbon Credits (Carbon Farming Initiative Act 2011). Canberra, Australia. www.comlaw.gov.au/Dssetails/C2011A00101.

Australian Industry Commission. (1992). The Australian Sugar Industry. Report No. 19. Australian Government Publishing Service, Canberra, Australia.

Australian Wet Tropics World Heritage Authority. (2007–2008). Annual Report 2007–2008. Wet Tropics Management Authority, Cairns, Queensland, Australia.

Australian Wet Tropics World Heritage Authority. (2011–2012). Annual Report 2011–2012. Wet Tropics Management Authority, Cairns, Queensland, Australia.

Bally, I. S. E., Lu, P. & Johnson, P. R. (2009). Mango breeding. In: Jain, S. M. & Priyadarshan, P. M. (Eds.), *Breeding Plantation Tree Crops, Tropical Species*, Chapter 2. Springer Science+Business Media, LLC, New York.

Banfai, D. S. & Bowman, D. M. J. S. (2005). Dynamics of a savanna-forest mosaic in the Australian monsoon tropics inferred from stand structures and terrestrial aerial photography. *Australian Journal of Botany* 53:185–194.

Barlow, B. A. & Hyland, B. P. M. (1988). The origins of the flora of Australia's wet tropics. *Proceedings of the Ecological Society of Australia* 15:1–17.

Barrett, D. (2011). Time scales and dynamics of carbon in Australia's savannas. In: Hill, M. J. & Hanan, N. P. (Eds.), *Ecosystem Function in Savannas: Measurement and Modelling at Landscape to Global Scales*, pp. 347–366. CRC Press, Boca Raton, FL, USA.

Barron, O. V., Crosbie, R. S., Charles, S. P. et al. (2011). Climate change impact on groundwater resources in Australia. Waterlines Report Series No. 67. National Water Commission, Commonwealth of Australia.

Bartlett, T. (2017). Promoting results and benefits arising from ACIAR's investment in international forestry research. *Australian Forestry* 80:119–120.

Bayliss, P., van Dam, R. A. & Bartolo, R. E. (2012). Quantitative ecological risk assessment of the Magela Creek floodplain in Kakadu National Park, Australia: Comparing point source risks from the Ranger Uranium Mine to diffuse landscape scale risks. *Human and Ecological Risk Assessment: An International Journal* 18:115–151.

Bellwood, D. R., Hoey. A. S., Ackerman, J. L. & Depczynski, M. (2006). Coral bleaching, reef fish community phase shifts and the resilience of coral reefs. *Global Change Biology* 12:1587–1594.

Belperio, A. P. (1979). Negative evidence for a mid-Holocene high sea level along the coastal plain of the Great Barrier Reef Province. *Marine Geology* 32:M1–M9.

Bender, D., Diaz-Pulido, G. & Dove, S. (2014). Warming and acidification promote cyanobacterial dominance in turf algal assemblages. *Marine Ecology Progress Series* 517:271–284.

Benson, S. M. & Cook, P. (2005). Underground geological storage. In: *Carbon Dioxide Capture: Special Report of the Intergovernmental Panel on Climate Change (IPCC)*, pp. 5–134. Cambridge University Press, Interlaken, Switzerland.

Beringer, J., Hutley, L. B., Tapper, N. J. & Cernusak, L. A. (2007). Savanna fires and their impact on net ecosystem productivity in North Australia. *Global Change Biology* 13:990–1004.

Berkelmans, R. (2002). Time-integrated thermal bleaching thresholds of reefs and their variations on the Great Barrier Reef. *Marine Ecology Progress Series* 229:73–82.

Berkelmans, R., De'ath, G. & Kininmonth, S. (2004). A comparison of the 1988 and 2002 coral bleaching events on the Great Barrier Reef; spatial correlation, patterns and predictions. *Coral Reefs* 23:74–83.

Berry, L., Taylor, A. R., Lucken, U. et al. (2002). Calcification and inorganic carbon acquisition in coccolithophores. *Functional Plant Biology* 29:289–299.

Berryman, C. A., Eamus, D. & Duff, G. A. (1994). Stomatal responses to a range of variables in two tropical tree species grown with CO_2 enrichment. *Journal of Experimental Botany* 45:539–546.

Bews, J. W. (1927). Studies in the ecological evolution of the angiosperms. *New Phytologist* 26:1–21.

Bird, M. I. & Pousal, P. (1997). Variations of $\delta_{13}C$ in the surface soil organic carbon pool. *Global Biogeochemical Cycles* 11:313–322.

Birkeland, C. & Lucas, J. S. (1990). *Acanthaster Planci: Major Management Problem of Coral Reefs*. CRC Press, Boca Raton, FL, USA.

Blach, I., Beckmann, C., Brown, G. P. & Shine, R. (2014). Effects of an invasive species on refuge-site selection by native fauna: The impact of cane toads on native frogs in the Australian tropics. *Austral Ecology* 39:50–59.

Black, C. C. & Osmond, C. B. (2003). Crassulacean acid metabolism photosynthesis: "working the night shift". *Photosynthesis Research* 76:329–341.

Blain, S., Quéquiner, B. & Wagener, T. (2007). Effect of natural iron fertilisation on carbon sequestration in the Southern Ocean. *Nature* 446:1070–1074.

Blais, A.-M., Lorrain, S., Plourde, Y. & Varfalvy, L. (2005). Organic carbon densities of soils and vegetation of tropical, temperate and boreal forests. In: Trembay, A., Varfalvy, L., Roehm, C. & Garneau, M. (Eds.), *Greenhouse Gas Emissions – Fluxes and Processes*, pp. 155–185. Springer-Verlag, Berlin, Germany.

Boland, K. T. (2008). Water quality profile and bathymetric survey of Whites and the Intermediate open cuts, Rum Jungle. Report by Tropical Water Solutions, Darwin, Northern Territory.

Bonan, G. B. (2008). Forests and climate change: Forcings, feedbacks and climate benefits of forests. *Science* 320:1444–1449.

Booth, D. J. & Beretta, G. A. (2002). Changes in a fish assemblage after a coral bleaching event. *Marine Ecology Progress Series* 245:205–212.

Borrell, A., Garside, A. & Fukai, S. (1997a). Improving efficiency of water use for irrigated rice in a semi-arid tropical environment. *Field Crops Research* 52:231–248.

Borrell, A. K., Garside, A. L., Fukai, S. & Reid, D. J. (1997b). Season and plant type affect the response of rice yield to nitrogen fertilization in a semi-arid tropical environment. *Australian Journal of Agricultural Research* 49:179–190.

Bosetti, V. & Lubowski, R. (Eds.). (2010). *Deforestation and Climate Change: Reducing Carbon Emissions from Deforestation and Forest Degradation*. Edward Elgar, Cheltenham, UK.

Bourne, A. & Norman, K. L. (1990). Burdekin irrigation field crops costs and returns: 1990/1. Queensland Department of Primary Industries Report No. RQM 90003, p. 50.

Bowman, D. M. J. S. (2001). Future eating and country keeping: What role has environmental history in the management of biodiversity. *Journal of Biogeography* 28:549–564.

Bowman, D. M. J. S. & Prior, L. D. (2005). Why do evergreen trees dominate the Australian seasonal tropics? *Australian Journal of Botany* 53:379–399.

Boyd, P. W., Jickells, T., Law, C. S. et al. (2007). Mesoscale iron enrichment experiments (1993–2005): Synthesis and future directions. *Science* 315:612–617.

Bradford, M. G., Metcalfe, D. J., Ford, A. et al. (2014). Floristics, stand structure and aboveground biomass of a 25-ha rainforest plot in the wet tropics of Australia. *Journal of Tropical Forest Science* 26:543–553.

Brand, J. E., Norris, L. J. & Dumbrell, I. C. (2012). Estimated heartwood weights and oil concentrations within 16-year-old Indian sandalwood (*Santalum album*) trees planted near Kununurra, Western Australia. *Australian Forestry* 75:225–232.

Bray, R. A. (1994). The Leucaena psyllid. In: Gutteridge, R. C. & Shelton, H. M. (Eds.), *Forage Tree Legumes in Tropical Agriculture*, pp. 283–291. CAB International, Wallingford, UK.

Brennan, F. (1998). *The Wik Debate: Its Impact on Aborigines, Pastoralists and Miners.* UNSW Press, Sydney, Australia.

Bridgeman, H. A. & Timms, B. V. (2012). Australia, climate and lakes. In: Bengtsson, L., Herschy, R. W. & Fairbridge, R.W. (Eds.), *Encyclopedia of Lakes and Reservoirs.* Springer, Dordrecht.

Brierley, A. S. & Kingsford, M. J. (2009). Impacts of climate change on marine organisms and ecosystems. *Current Biology* 19:R602–R614.

Brodie, J., Fabricius, K., De'ath, G. & Okaji, K. (2005). Are increased nutrient inputs responsible for more outbreaks of crown-of-thorns starfish? An appraisal of the evidence. *Marine Pollution Bulletin* 51:266–278.

Brodie, J. E., McKergow, L. A., Prosser, I. P. et al. (2003). Sources of sediment and nutrients exported to the Great Barrier Reef World Heritage Area. Australian Centre for Tropical Freshwater Research. Report No. 03/11. James Cook University, Townsville.

Brooker, M. I. H. & Kleinig, D. A. (2004). *Field Guide to Eucalypts.* Vol. 3. *Northern Australia,* 2nd ed. Bloomings Books, Melbourne, Australia.

Broughton, K. J., Smith, R. A., Duursma, R. A. et al. (2017). Warming alters the positive impact of elevated CO_2 concentrations on cotton growth. *Functional Plant Biology* 44:267–278.

Brown, B. (1999). Occurrence and impact of *Phytophthora cinnamoni* and other *Phytophthora* species in forests of the Wet Tropics World Heritage area and of the Mackay region, Queensland. In: Gadek, P. A. (Ed.), *Patch Death in Tropical Queensland Rainforest Associations,* pp. 41–76. CRC Tropical Forest Ecology and Management, Cairns, Queensland, Australia.

Brown, B. E. & Bythell, J. C. (2005). Perspectives on mucus secretion in reef corals. *Marine Ecology Progress Series* 296:291–309.

Brown, S. A., Iverson, L. R. & Lugo, A. E. (1992b). Land use and biomass changes in Peninsular Malaysia during 1972–82: A GIS approach. In: Dale, V. H. (Ed.), *Effects of Land-Use Change on Atmospheric CO_2 Concentrations, South East Asia Case-Study.* Springer-Verlag, New York, USA.

Brown, S. A., Lugo, A. E. & Iverson, L. R. (1992a). Processes and lands for sequestering carbon in the tropical forest landscape. *Water, Air and Soil Pollution* 64:139–155.

Bruce, S. L., Noller, B. N., Grigg, A. H. et al. (2003). A field study conducted at Kidston Gold Mine to evaluate the impact of arsenic and zinc from mine tailing to grazing cattle. *Toxicology Letters* 137:23–34.

Buck, B. H., Rosenthal, H. & Saint-Paul, U. (2002). Effect of increased irradiance on thermal stress on the symbiosis of *Symbiodinium microadriaticum* and *Tridacna gigas. Aquatic Living Resources* 15:107–117.

Bunt, J. S. (1982). Studies of mangrove litter fall in tropical Australia. In: Clough, B. F. (Ed.), *Mangrove Ecosystems in Australia: Structure, Function and Management,* pp. 223–237. Australian Institute of Marine Science.

Burbidge, N. T. (1960). The phytogeography of the Australian region. *Australian Journal of Botany* 8:75–211.

Burford, M., Revill, A., Palmer, D. et al. (2011). River regulation alters drivers of primary productivity along a tropical river-estuary system. *Marine and Freshwater Research* 62:141–151.

Bussell, W. T. & Bonin, M. J. (1999). Study of asparagus production in Western Samoa. *New Zealand Journal of Crop and Horticultural Science* 27:163–168.

Bussell, W. T., Olsen, J. K., Robinson, C. & Bright, J. P. (2002). Asparagus in tropical Australia – The first fifteen years. *Australian Journal of Agricultural Research* 53:729–736.

Butler, A. J. & Jernakoff, P. (1999). *Seagrass in Australia: Strategic Review and Development of an R & D Plan.* CSIRO Publishing, Melbourne, Victoria, Australia.

Butler, D. W. & Fairfax, R. J. (2003). Buffel grass and fire in a gidgee and brigalow woodland: A case study from central Queensland. *Ecological Management & Restoration* 4:120–125. Queensland Herbarium, Brisbane, Queensland, Australia.

Calvin, M. (1989). Forty years of photosynthesis and related activities. *Photosynthesis Research* 21:3016.

Carberry, P. S., Muchow, R. C. & McCown, R. L. (1993). A simulation model of kenaf for assisting fibre industry planning in northern Australia IV. Analysis of climate risk. *Australian Journal of Agricultural Research* 44:713–730.

Carbone, M. S., Czimczik, C. I. & Keenan, T. F. (2013). Age, allocation and availability of non-structural carbon in mature red maple trees. *New Phytologist* 200:1145–1155.

Carruthers, I. B., Griffiths, D. J., Home, V. & Williams, L. R. (1984). Hydrocarbons from *Calotropis procera* in Northern Australia. *Biomass* 4:275–282.

Carruthers, S. (2015). Protected cropping in the tropics. *Practical Hydroponics & Greenhouses* 152:36–41.

Catterall, C. P. & Harrison, D. A. (2006). Rainforest restoration activities in Australia's tropics and sub-tropics. Co-operative Research Centre for Tropical Rainforest Ecology and Management, Special Report.

Cermeño, P., Dutkiewicz, S., Harris, R. P. et al. (2008). The role of nutricline depth in regulating the ocean carbon cycle. *Proceedings of the National Academy of Sciences of the United States of America* 105:20344–20349.

Cernusak, L. A., Winter, K., Dalling, J. W. et al. (2013). Tropical forest responses to increasing atmospheric CO_2: Current knowledge and opportunities for future research. *Functional Plant Biology* 40:531–551.

Charles, L. S., Dwyer, J. M. & Mayfield, M. M. (2017). Rainforest seed rain into abandoned tropical Australian pasture is dependent on adjacent rainforest structure and extent. *Austral Ecology* 42:238–249.

Chase, A. (2009). *Pama Malngkana*. In: Anderson, S. (Ed.), *Pelletier: The Forgotten Castaway of Cape York*. Melbourne Books, Melbourne, Australia.

Chauhan, Y. S., Thorburn, P., Biggs, J. S. & Wright, G. C. (2015). Agronomic benefits and risks associated with the irrigated peanut–maize production system under a changing climate in northern Australia. *Crop and Pasture Science* 66:1167–1179.

Chave, J., Andalo, C., Brown, S. et al. (2005). Tree allometry and improved estimation of carbon stocks and balance in tropical forests. *Oecologia* 145:87–99.

Chen, X., Hutley, L. B. & Eamus, D. (2003). Carbon balance of a tropical savanna of northern Australia. *Oecologia* 137:405–416.

Chen, X., Hutley, L. B. & Eamus, D. (2005). Soil organic content at a range of northern Australian tropical savannas with contrasting site histories. *Plant & Soil* 268:161–171.

Chippendale, G. M. & Murray, L. R. (1963). Northern Territory Administration, Animal Industries Branch, Extension Article No. 2:85.

Chitdeshwari, T., Bell, R. W., Anderson, J. & Phillips, I. R. (2012). Plant-available manganese in bauxite residue sand amended with compost and residue mud. *Soil Research* 50:416–423.

Choat, B., Ball, M. C., Luly, J. G., Donnelly, C. F. & Holtum, J. A. M. (2006). Seasonal patterns of leaf gas exchange and water relations in dry rain forest trees of contrasting leaf phenology. *Tree Physiology* 26:657–664.

Christopher, D. M., Harper, R. J. & Keenan, R. J. (2012). Current status and future prospects for carbon forestry in Australia. *Australian Forestry* 75:200–212.

Church, J. A. (1987). East Australian current adjacent to the Great Barrier Reef. *Australian Journal of Marine and Freshwater Research* 38:671–683.

Clark, R., Gonzales, D., Dillon, P. et al. (2015). Reliability of water supply from stormwater harvesting and managed aquifer recharge in an urbanizing catchment and climate change. *Environmental Modelling & Software* 72:117–125.

Clarke, P. J., Latz, P. K. & Albrecht, D. E. (2005). Long-term changes in semi-arid vegetation: Invasion of an exotic perennial grass has larger effects than rainfall variability. *Journal of Vegetation Science* 16:237–248.

Clem, R. L. (2004). Animal production from legume-based ley pastures in south-eastern Queensland. In: Whitbread, A. M. & Pengelly, B. C. (Eds.), *Tropical Legumes for Sustainable Farming in Southern Africa and Australia*, pp. 136–144. Australian Centre for International Agricultural Research, Canberra, Australia.

Clode, P. L. & Marshall, A. T. (2002). Low temperature FESEM of the calcifying interface of a scleractinian coral. *Tissue & Cell* 34:187–198.

Cogo, K. (2010). Sugar cane makes way for wetland wildlife in Queensland's Burdekin. *Wetlands Australia* 18:46.

Cohen, A. L. & McConnaughty, T. A. (2003). Geochemical perspectives on coral mineralization. *Reviews of Mineralogy & Geochemistry* 54:151–187.

Coles, R. G., Lee Long, W. J., Squire, B. A. et al. (1987). Distribution of seagrasses and associated juvenile commercial penaeid prawns in north-eastern Queensland waters. *Australian Journal of Marine and Freshwater Research* 38:103–119.

Comeau, S., Edmunds, P. J., Spindel, N. B. et al. (2013). The response of eight coral reef calcifiers to increasing partial pressure of CO_2 do not exhibit a tipping point. *Limnology and Oceanography* 58:388–398.

Comerford, E., Norman, P. L. & Le Grand, J. (2015). Is carbon forestry viable? A case study for Queensland Australia. *Australian Forestry* 78:169–179.

Congdon, R. A. & Herbohn, J. L. (1993). Ecosystem dynamics of disturbed and undisturbed sites in north Queensland wet tropical rain forest. I. Floristic composition, climate and soil chemistry. *Journal of Tropical Ecology* 9:349–363.

Cook, F. W. (1991). Mining at Coronation Hill. Background paper prepared for the Department of Science, Technology & Environment, Resource Assessment Commission, Australian Government.

Cook, G. D., Leidloff, A. C., Eager, R. W. et al. (2005). The estimation of carbon budgets of frequently burnt tree stands in savannas of northern Australia, using allometric analysis and isotope discrimination. *Australian Journal of Botany* 53:621–630.

Copland, J. W. & Lucas, J. S. (Eds.). (1988). *Giant Clams in Asia and the Pacific*. Proceedings of the Conference held at James Cook University, Townsville, ACIAR, Canberra, Australia.

Corbin, K. R., Byrt, C. S., Bauer, S. et al. (2015). Prospecting for energy-rich renewable raw materials: Agave leaf case study. *PLoS One* 10:e0135382.

Couper, P. J., Wilmer, J. W., Roberts, L. et al. (2005). Skinks currently assigned to *Carlia aerate* (Scincidae: Lygosominae) of north-eastern Queensland: A preliminary study of cryptic diversity and two new species. *Australian Journal of Zoology* 53:35–49.

Cowie, I. D., Armstrong, M. D., Woinarski, J. C. Z. et al. (2000). An overview of the floodplains. In: Cowie, I. D., Short, P. S. & Osterkamp, M. M. (Eds.), *Floodplain Flora: A Flora of the Coastal Floodplains of the Northern Territory*. Australian Biological Resources Study, Canberra, Australia.

Cramer, W., Bondeau, A., Schapoff, S. et al. (2004). Tropical forests and global carbon cycle: Impacts of atmospheric carbon dioxide, climate change and deforestation. *Philosophical Transactions of the Royal Society B: Biological Sciences* 359:331–343.

Crayn, D. M., Winter, K., Schulte, K. & Smith, J. A. C. (2015). Photosynthetic pathways in Bromeliaceae: Phylogenetic and ecological significance of CAM and C-3 based on carbon isotope ratios for 1,893 species. *Botanical Journal of the Linnean Society* 178:169–221.

Crouzeilles, R., Ferreira, M., Chazdon, R. L. et al. (2017). Ecological restoration success is higher for natural regeneration than for active restoration in tropical forests. *Science Advances* 3:e1701345.

Crowley, G. M. & Garnett, S. T. (2000). Changing fire management in the pastoral lands of Cape York Peninsula of north-east Australia. *Australian Geographic Studies* 38:10–26.

Cullen, B. C. & Hill, J. H. (2005). A survey of the use of lucerne, butterfly pea and lablab in ley pastures in the mixed farming systems of northern Australia. *Tropical Grasslands* 39:24–32.

Cunningham, S. C. & Read, J. (2003). Do temperate rainforest trees have a greater ability to acclimate to changing temperatures than tropical rainforest trees. *New Phytologist* 157:55–64.

Damayanti, F., Lawn, R. J. & Bielig, L. M. (2010). Genotypic variation in domesticated and wild accessions of the tuberous legume *Vigna vexillata* (L) A. Rich. *Crop & Pasture Science* 61:771–784.

Daughty, C. E., Metcalfe, D. B., da Costa, M. C. et al. (2014). The production, allocation and cycling of carbon in a forest on fertile *terra preta* soil in eastern Amazonia compared with a forest on adjacent infertile soil. *Plant Ecology & Diversity* 7:41–53.

Davidson, B. R. (1965). *The Northern Myth: A Study of the Physical and Economic Limits to Agriculture and Pastoral Development in Tropical Australia.* Melbourne University Press, Carlton, Victoria, Australia.

Davies, A. P., Pufke, U. S. & Zalucki, M. P. (2009). *Trichogramma* (Hymenoptera: Trichogrammatidae) ecology in the tropical Bt transgenic cotton cropping system: Sampling to improve seasonal pest impact estimates in the Ord River Irrigation Area, Australia. *Journal of Economic Entomology* 102:1018–1031.

DePaolo, D. J. & Cole, D. R. (2013). Geochemistry of geological carbon sequestration: An overview. In: DePaolo, D. J., Cole, D. R., Navrotsky, A. & Bourg, I. C. (Eds.), *Geochemistry of Geologic CO$_2$ Sequestration (Reviews in Mineralogy and Geochemistry* 77:1–14).

Department of Mines and Petroleum and Environmental Protection Authority-Western Australia (DMPEPA), Western Australia. (2015). Guidelines for preparing mine-closure plans. Government of Western Australia. http://www.dmp.wa.gov.au/documents/envir onment/ENV-MEB-121-pdf

De Souza, A. P., Gaspar, M., da Silva, E. A. et al. (2008). Elevated CO$_2$ increases photosynthesis, biomass and productivity and modifies gene expression in sugarcane. *Plant Cell and Environment* 31:1116–1127.

De Souza Dias, M. O., Filho, R. M., Mantelatto, P. E. et al. (2015). Sugarcane processing for ethanol and sugar in Brazil. *Environmental Development* 15:35–51.

Depczynski, M., Gilmour, J. P., Ridgway, T. et al. (2013). Bleaching, coral mortality and subsequent survivorship on a West Australian fringing reef. *Coral Reefs* 32:233–238.

Dinar, A. & Mendelsohn, R. (Eds.). (2011). *Handbook on Climate Change and Agriculture.* Edward Elgar, Cheltenham, UK.

D'Odorico, P. & Bhattacham, A. (2012). Hydrological variability in dryland regions: Impacts on ecosystem dynamics and food security. *Philosophical Transactions of the Royal Society B: Biological Sciences* 367:3145–3157.

Done, A. C. & Wood, I. M. (1981). Genetic variation and cross pollination in kenaf and roselle. In: Wood, I. M. & Stewart, G. A. (Eds.), *Kenaf as a Potential Source of Pulp in Australia*, pp. 29–35. Proceedings of Kenaf Conference, CSIRO, Brisbane, Queensland, Australia.

Drake, P. L. & Franks, P. J. (2003). Water resource partitioning, stem xylem hydraulic properties and plant water-use strategies in a seasonally dry riparian tropical rainforest. *Oecologia* 137:321–329.

Drew, E. A. (1983). *Halimeda* biomass, growth rates and sediment generation on reefs in the central great barrier reef province. *Coral Reefs* 2:101–110.

Eamus, D. (1999). Ecophysiological traits of deciduous and evergreen woody species in the seasonally dry tropics. *Trends in Ecology and Evolution* 14:11–16.

Eamus, D. & Prior, L. (2001). Ecophysiology of trees of seasonally dry tropics: Comparisons among phenologies. *Advances in Ecological Research* 32:113–197.

Eastick, R. J & Hearnden, M. N. (2006). Potential for weediness of Bt cotton in northern Australia. *Weed Science* 54:1142–1151.

EDO Qld. (2016). Laws passed to protect Queensland taxpayers against costly environmental clean-ups. Environmental Defence Office Report. (Update April 2016.)

Edwards, A., Hauser, P., Anderson, M. et al. (2001). A tale of two parks: Contemporary fire regimes of Litchfield and Nitmiluk National Parks, monsoonal northern Australia. *International Journal of Wildland Fire* 10:79–89.

Edwards, G. (2014). Stewardship payments: Remedy for drought pain. *Queensland Country Life* 13:15.

Edwards, W., Liddell, M. J., Franks, P. et al. (2017). Seasonal patterns in rainforest litterfall: Detecting endogenous and environmental influences from long-term sampling. *Austral Ecology* 43. doi:10.1111/acc.12559.

Eliot, I., Finlayson, C. M. & Waterman, P. (1999). Predicted climate change, sea-level rise and wetland management in the Australian wet-dry tropics. *Wetlands Ecology & Management* 7:63–81.

Ellison, J. C. & Stoddart, D. R. (1991). Mangrove ecosystem collapse during predicted sea-level rise. Holocene analogues and implications. *Journal of Coastal Research* 7:151–165.

Endean, R. (1969). Report on investigations made into aspects of the current *Acanthaster planci* (Crown-of-thorns) infestations of certain reefs of the Great Barrier Reef. Fisheries Branch, Queensland Department of Primary Industries, Brisbane, Queensland, Australia.

Erftemeijer, P. L. A., Wylie, N. & Hooper, G. J. (2018). Successful mangrove establishment along an artificially created creek at Port Hedland, Western Australia. *Marine and Freshwater Research* 69:134–143.

Erskine, P. D. (2002). Land clearing and forest rehabilitation in the wet tropics of north Queensland, Australia. *Ecological Management and Restoration* 3:136–138.

Erskine, P. D. (2004). Production versus rainforest diversity: Trade-offs or synergies in farm forestry systems? In: Erskine, P. D. & Catterall, C. P. (Eds.), Workshop Proceedings, Rainforest Ecology and Management CRC, Cairns, Queensland, Australia.

Erskine, P. D., Lamb, D. & Bristow, M. (2006). Tree species diversity and ecosystem function: Can tropical multi-species plantations generate greater productivity? *Forest Ecology & Management*, 233:205–210.

Esparon, M., Stoeki, N. & Gyuris, E. (2013). ECO certification in Queensland's Wet Tropics World Heritage Area: Is it good for business? In: Tisdell, C. A. (Ed.), *Handbook of Tourism Economics*, pp. 845–869. World Scientific, Singapore.

Evans, S. E., Byrne, K. M., Lauenroth, W. K. & Burke, I. C. (2011). Defining the limit to resistance in a drought-tolerant grassland: Long-term severe drought significantly reduces the dominant species and increases ruderals. *Journal of Ecology* 99:1500–1507.

Everist, S. L. (1974). *Poisonous Plants of Australia*, 2nd ed. Angus & Robertson, Sydney, Australia.

Everist, S. L. (1981). *Poisonous Plants of Australia*. Angus & Robertson, Sydney, Australia

Faith, D. P., Dostine, P. L. & Humphrey, C. L. (1995). Detection of mining impacts on aquatic macroinvertebrate communities: Results of a disturbance experiment and the design of a multivariate BACIP monitoring programme at Coronation Hill, Northern Territory. *Australian Journal of Ecology* 20:167–180.

Faithful, J. W. & Griffiths, D. J. (2000). Turbid flow through a tropical reservoir (Lake Dalrymple, Queensland, Australia): Response to a summer storm event. *Lakes and Reservoirs, Research and Management* 5:231–247.

Farrell, T. P. (1978). The spread and control of *Salvinia molesta* in Lake Moondarra, Mount Isa, Queensland. In: *Proceedings of 1st Conference of the Council of Australian Weed Science Societies*, Melbourne, Australia, April 1978, pp. 179–187.

Farrell, T. P., Finlayson, C. M. & Griffiths, D. J. (1979). Studies of the hydrobiology of a tropical lake in north-western Queensland. I. Seasonal changes in chemical characteristics. *Australian Journal of Marine and Freshwater Research* 30:579–595.

Feary, S., Kanowski, P., Altman, J. & Baker, R. (2010). Managing forest country: Aboriginal Australians and the forest sector. *Australian Forestry* 73:126–134.

Fensham, R. J. (2012). Fire regimes in Australian tropical savanna: Perspectives, paradigms and paradoxes. In: Bradstock, R. A., Gill, A. M. & Williams, R. J. (Eds.), *Flammable Australia: Regimes, Biodiversity and Ecosystems in a Changing World*, pp. 173–193. CSIRO Publishing, Collingwood, Victoria, Australia.

Fensham, R. J. & Fairfax, R. J. (2005). Preliminary assessment of gidgee (*Acacia cambagei*) woodland thickening in the Longreach district of Queensland. *The Rangeland Journal* 27:159–168.

Fensham, R. J. & Fairfax, R. J. (2007). Drought related tree death of savanna eucalypts: Species susceptibility, soil conditions and root architecture. *Journal of Vegetation Science* 18:71–80.

Fensham, R. J., Freeman, M. E., Laffineur, B. et al. (2017). Variable rainfall has a greater effect than fire on the demography of the dominant tree in a semi-arid *Eucalyptus* savanna. *Austral Ecology* 42:772–782.

Fernando, N., Manalil, S., Florentine, S. K. et al. (2016). Glyphosate resistance of C-3 and C-4 weeds under rising atmospheric CO_2. *Frontiers in Plant Science* 7:1–10.

Finlayson, C. M. (2005). Plant ecology of Australia's tropical wetlands: A review. *Annals of Botany* 96:541–555.

Finlayson, C. M., Farrell T. P. & Griffiths, D. J. (1984a). The hydrobiology of five man-made lakes in north-western Queensland. *Proceedings of the Royal Society of Queensland* 95:29–40.

Finlayson, C. M., Farrell, T. P. & Griffiths, D. J, (1984b). Studies of the hydrobiology of a tropical lake in north-eastern Queensland III. Growth, chemical composition and potential for harvesting of the aquatic vegetation. *Australian Journal of Marine and Freshwater Research* 35:525–536.

Flint, E. P. & Richards, J. F. (1991). Historical analysis of changes in land use and carbon stocks of vegetation in South and Southeast Asia. *Canadian Journal of Forest Research* 21:91–110.

Flint, N., Pearson, R. G. & Crossland, M. R. (2018). Reproduction and embryo viability of a range-limited tropical freshwater fish exposed to fluctuating hypoxia. *Marine and Freshwater Research* 69:267–276.

Francis, W. D. (1981). *Australian Rain-Forest Trees*. Australian Government Publishing Service, Canberra.

Franklin, D. C., Brocklehurst, P. S., Lynch, D. & Bowman, D. M. J. S. (2007). Niche differentiation and regeneration in the seasonally flooded *Melaleuca* forests of northern Australia. *Journal of Tropical Ecology* 23:457–467.

Franks, P. J., Adams, M. A., Amthor, J. S. et al. (2013). Sensitivity of plants to changing atmospheric CO_2 concentration: From the geological past to the next century. *New Phytologist* 197:1077–1094.

Frederick, S. E. &Newcomb, E. H. (1971). Ultrastructure and distribution of microbodies in leaves of grasses with and without CO_2 – Photorespiration. *Planta* 96:152–174.

Freeman, M. E., Vesk, P. A., Murphy, B. P. et al. (2017). Defining the fire trap: Extension of the persistence equilibrium model in mesic savannas. *Austral Ecology*. doi:10.1111/aec.12516.

Freiberg, M. & Turton, S. M. (2007). Importance of drought on the distribution of the bird's nest fern, *Asplenium nidus*, in the canopy of a lowland tropical rainforest in north-eastern Australia. *Austral Ecology* 32:70–76.

Galbraith, D., Malhi, Y., Affum-Baggoe, K. et al. (2013). Residence times of woody biomass in tropical forests. *Plant Ecology & Diversity* 6:139–157.

Galloway, R. W. (1982). Distribution and physiographic patterns of Australian mangroves. In: Clough, B. F. (Ed.), *Mangrove Ecosystems in Australia: Structure, Function and Management*, pp. 31–54. Australian Institute of Marine Sciences.

Gammage, B. (2011). *The Biggest Estate on Earth: How Aborigines Made Australia*. Allen & Unwin, Crows Nest, New South Wales, Australia.

Garcia-Flores, R., Higgins, A., Prestwidge, D. & McFallan, S. (2014). Optimum location of spelling yards for the northern Australian beef supply chain. *Computers & Electronics in Agriculture* 102:134–145.

Geddes, P. (2016). A review of damage caused to Australian forestry plantations by tropical cyclones during the period from 1998 to 2015. *Australian Forest Grower* 39:32–33.

Ghannoum, O., Phillips, N. G., Conroy, J. P. et al. (2010). Exposure to preindustrial, current and future atmospheric CO_2 and temperature differentially affects growth and photosynthesis in *Eucalyptus*. *Global Change Biology* 16:303–319.

Gilbert, M. A., Edwards, D. G., Shaw, K. A. & Jones, R. K. (1989). Effect of phosphorus supply on three perennial *Stylosanthes* species in tropical Australia. II. Phosphorus and Nitrogen within the plant and implications for grazing animals. *Australian Journal of Agricultural Research* 40:1205–1216.

Gillard, P. (1979). Improvement of native pastures with Townsville stylo in the dry tropics of sub-coastal northern Queensland. *Australian Journal of Experimental Agriculture and Animal Husbandry* 19:325–336.

Gillespie, S. & van den Bold, M. (2017). Agriculture, food systems and nutrition: Meeting the challenge. *Global Challenges*. doi:10.1002/gch2.201600002.

Gleason, S. M., Williams, L. J. Read, J. et al. (2008). Cyclone effects on the structure and production of a tropical upland rainforest: Implications for life-history trade-offs. *Ecosystems* 11:1277–1290.

Gleeson, T., Martin, P. & Mifsud, C. (2012). Northern Australian beef industry: Assessment of risks and opportunities. Australian Government, Department of Agriculture Fisheries and Food.

Gleixner, G. (2013). Soil organic matter dynamics: A biological perspective derived from the use of compound-specific isotopes. *Ecological Research* 28:683–695.

Goldberg, J., Birtles, A., Marshall, N. et al. (2018). The role of Great Barrier Reef tourism operators in addressing climate change through strategic communication and direct action. *Journal of Sustainable Tourism* 26:238–256.

Gould, S. F. (2010). Does post-mining rehabilitation on the Weipa bauxite plateau restore bird habitat values? PhD thesis, Australian National University, Canberra, Australia.

Gourley, C. J. P., Aarons, S. R. & Powell, J. M. (2012). Nitrogen use efficiency and manure management practices in contrasting dairy production systems. *Agriculture, Ecosystems and Environment* 147:73–81.

Grace, P. R. & Basse, B. (2012). Offsetting gas emissions through biological carbon sequestration in North-Eastern Australia. *Agricultural Systems* 105:1–6.

Graetz, G. (2015a). Ranger uranium mine and the Mirarr (Part 1), 1970–2000: The risk of "riding roughshod". *The Extractive Industries and Society* 2:132–141.

Graetz, G. (2015b). Ranger uranium mine and the Mirarr (Part 2). 2000–2014: "A risk to them is a risk to us". *The Extractive Industries and Society* 2:142–152.

Graetz, R. D. & Skjemstad, J. O. (2003). The charcoal sink of biomass burning on the Australian continent. CSIRO Atmospheric Research, Technical Paper No. 64.

Grice, A. M. (1999). Studies on the giant clam – Zooxanthellae symbiosis. Thesis, James Cook University, Townsville, Australia, p. 164.

Grice, A. M. & Bell, J. D. (1999). Application of ammonium to enhance the growth of giant clams (*Tridacna maxima*) in the land-based nursery: Effects of size-class, stocking density and nutrient concentration. *Aquaculture* 170:17–28.

Griffiths, D. J. (2016). *Freshwater Resources of the Tropical North of Australia: A Hydrobiological Perspective.* Nova Science Publishers, Inc., New York, USA.

Griffiths, D. J. & Saker, M. L. (2003). The Palm Island mystery disease 20 years on: A review of research on the cyanotoxin Cylindrospermopsin. *Environmental Toxicology* 18:78–93.

Grundy, M. J., Bryan, B. A., Nolan, M., Battaglia, M. et al. (2016). Scenarios for Australian agricultural production and land use to 2050. *Agricultural Systems* 142:70–83.

Halford, A., Cheal, A. J., Ryan, D. et al. (2004). Resilience to large-scale disturbance in coral and fish assemblages on the Great Barrier Reef. *Ecology* 85:1892–1905.

Hart, A. M., Klumpp, D. W. & Russ, G. R. (1996). Response of herbivorous fishes to crown-of-thorns starfish *Acanthaster planci* outbreaks. II. Density and biomass of selected species of herbivorous fish and fish habitat correlations. *Marine Ecology Progress Series* 132:21–30.

Hasagawa, T., Sakai, H., Tokida, T. et al. (2013). Rice cultivar responses to elevated CO_2 at two free-air CO_2 enrichment (FACE) sites in Japan. *Functional Plant Biology* 40:148–159.

Hatch, M. D. & Slack, C. R. (1967). Further studies on a new pathway of photosynthetic carbon dioxide fixation in sugar-cane and its occurrence in other plant species. *Biochemical Journal* 102:417–422.

Havemann, P., Thiret, D., Marsh, H. & Jones, C. (2005). Traditional use of marine resources agreements and dugong hunting in the Great Barrier Reef World Heritage Area. *Environmental and Planning Law Journal* 22:258–280.

Hawkins, P. R. & Griffiths, D. J. (1993). Artificial destratification of a small tropical reservoir: Effects upon the phytoplankton. *Hydrobiologia* 254:169–181.

Hawkins, P. R., Runnegar, M. T. C., Jackson, A. R. B. & Falconer, I. R. (1985). Severe hepatotoxicity caused by a tropical cyanobacterium *Cylindrospermopsis raciborskii* (Woloszynska) Seenaya and Subba Raju isolated from a domestic supply reservoir. *Applied and Environmental Microbiology* 50:1292–1295.

Hendrey, G. R., Elsworth, D. S., Lewin, K. F. & Nagy, J. (1999). A free-air enrichment system for exposing tall forest vegetation to elevated atmospheric CO_2. *Global Change Biology* 5:293–309.

Henry, G. W. & Lyle, J. M. C. (2003). The National Recreational and Indigenous Fishing Survey. Fisheries Research and Development Corporation Project 99/158.

Herbohn, J. L., Harrison, S. R. & Emtage, N. F. (1999). Potential performance of rainforest and eucalypt timber in plantations in North Queensland. *Australian Forestry* 62:79–87.

Hettiarachchi, D. S., Liu, Y., Jose, S. et al. (2012). Assessment of Western Australian sandalwood seeds for seed oil production. *Australian Forestry* 75:246–250.

Hewson, H. J. & George, A. S. (1984). Santalaceae. In: George, A. S. (Ed.), *Flora of Australia*, Vol. 22, pp. 29–67. Bureau of Flora and Fauna, Australian Government Publishing Service, Canberra, Australia.

Higgins, E., Walton, I., Chilcott, C. et al. (2012). A framework for optimizing capital investment and operations in livestock logistics. *The Rangeland Journal* 35:181–191.

Hilbert, D. W., Ostendorf, B. & Hopkins, M. S. (2001). Sensitivity of tropical forests to climate change in the humid tropics of north Queensland. *Austral Ecology* 26:590–603.

Hill, R. S. (1994). *History of Australian Vegetation, Cretaceous to Recent.* Cambridge University Press, Cambridge, UK.

Hoegh-Guldberg, O. & Bruno, J. F. (2010). The impact of climate change on the world's marine ecosystems. *Science* 328:1523.

Holm, A. M. (1973). The effect of high temperature pre-treatments on germination of Townville stylo seed material. *Australian Journal of Experimental Agriculture and Animal Husbandry* 13:190–192.

Holtum, J. A. M., Chambers, D., Morgan, T. & Tan, D. K. Y. (2011). Agave as a biofuel feedstock in Australia. *Global Change Biology Bioenergy* 3:58–67.

Holtum, J. A. M. & Winter, K. (1999). Degrees of crassulacean acid metabolism in tropical epiphytic and lithophytic ferns. *Australian Journal of Plant Physiology* 26:749–757.

Holtum, J. A. M. & Winter, K. (2001). Are plants growing close to the floors of tropical forests exposed to markedly elevated concentrations of carbon dioxide? *Australian Journal of Botany* 49:629–636.

Holtum, J. A. M. & Winter, K. (2003). Photosynthetic CO_2 uptake in seedlings of two tropical tree species exposed to oscillating elevated concentrations of CO_2. *Planta* 218:152–158.

Holtum, J. A. M. & Winter, K. (2010). Elevated [CO_2] and forest vegetation: More a water issue than a carbon issue. *Functional Plant Biology* 37:694–702.

Holtum, J. A. M. & Winter, K. (2014). Limited photosynthetic plasticity in the leaf-succulent CAM plant *Agave angustifolia* grown at different temperatures. *Functional Plant Biology* 41:843–849.

Horridge, M., Madden, J. & Wittwer, G. (2005). The impact of the 2002–03 drought on Australia. *Journal of Policy Modelling* 27:285–308.

Horton, D. R. (1982). The burning question: Aborigines, fire, and Australian ecosystems. *Australian Journal of Anthropology* 13:237–252.

Horton, H. (1976). *Around Mount Isa: A Guide to the Flora and Fauna*. University of Queensland Press, St. Lucia, Queensland, Australia.

Houghton, R. A., Boone, R. D., Fruci, J. R. et al. (1987). The flux of carbon from terrestrial ecosystems to the atmosphere in 1980 due to changes in land use: Geographic distribution of the global flux. *Tellus* 39B:122–139.

Houghton, R. A., Byers, B. & Nassikas, A. A. (2015). A role for tropical forests in stabilizing atmospheric CO_2. *Nature Climate Change* 5:1022–1023.

Houston, W. A., Melzer, A. & Black, R. L. (2018). Recovery of reptile, amphibian and mammal assemblages in industrial post-mining landscapes following open-cut coal mining. *Proceedings of the Royal Society of Queensland* 123:31–47.

Hu, J., Herbohn, J., Chazdon, R. L. et al. (2018). Recovery of species composition over 46 years in a logged Australian tropical forest following different intensity silvicultural treatments. *Forest Ecology and Management* 409:660–666.

Humphrey, C. L., Thurtell, L., Pidgeon, R. W. J. et al. (1998). A model for assessing the health of Kakadu's streams. In: Vardon, M. & Noske, R. (Eds.), *Biology in the Wet-Dry Tropics: Still Wet Behind the Ears?* Proceedings of the Symposium of the Australian Institute of Biology, Northern Territory University, Darwin, Northern Territory, Australia, 11 July (*Australian Biologist* 12(1):33–42).

Humphries, S. E., Groves, R. H. & Mitchell, D. S. (1991). Plant invasions of Australian ecosystems: A status review and management directions. In: Longmore, R. (Ed.), *Plant Invasions: The Incidence of Environmental Weeds in Australia*, Vol. 2, pp. 1–134.

Hutton, E. M. & Gray, S. G. (1959). Problems in adapting *Leucaena glauca* as a forage for the Australian tropics. *The Empire Journal of Experimental Agriculture* 29:187–196.

Hynes, R. A. & Chase, A. K. (1982). Plants, sites and domiculture: Aboriginal influences upon plant communities in Cape York Peninsula. In: Powell, J. & White J. P. (Eds.), *Plants and People. Archaeology in Oceania*, 17:38–50.

Hynes, R. A. & Panetta, F. D. (1994). Pest invasion, land sustainability and the maintenance of biodiversity. *Australian Biologist* 7:4–22.

Indigo, N., Smith, J., Webb, J. K. & Phillips, B. (2018). Not such silly sausages: Evidence suggests northern quolls exhibit aversion to toads after training with toad sausages. *Austral Ecology* 43. doi:10.1111/acc.12595.

IPCC. (2007). Climate change 2007: The physical science basis. Contribution of Working Group 1 to the Fourth Assessment Report of the Intergovernmental Panel on Climate Change, Cambridge University Press.

Jeffree, R. A., Twining, J. R. & Thomson, J. (2001). Recovery of fish communities in the Finniss River, Northern Australia following remediation of the Rum Jungle Uranium/ Copper mine site. *Environmental Science & Technology* 35:2932–2941.

Jensen, F., Nielsen, F. & Nielsen, R. (2014). Increased competition for aquaculture from fisheries. Does improved fisheries management limit aquaculture growth? *Fisheries Research* 159:25–33.

Johnson, G. I. (2008). Status of mango postharvest disease management R & D: Options and solutions for the Australian mango industry. Horticulture Australia Final Report for project MG08017:1–130.

Jones, P. (2002). Estimating returns on plantation-grown sandalwood (*Santalum spicatum*). *Sandalwood Information Sheet*. Forest Products Commission, Perth, Western Australia, Australia.

Jones, R. (1975). The Neolithic, Palaeolithic and the hunting gardeners: Man and land in the antipodes. In: Suggate, R. P. & Cresswell, M. M. (Eds.), *Quaternary Studies*, pp. 21–34. The Royal Society of New Zealand, Wellington, New Zealand.

Jones, R. J. & Megarrity, R. G. (1986). Successful transfer of DHP-degrading bacteria from Hawaiian goats to Australian ruminants to overcome the toxicity of leucaena. *Australian Veterinary Journal* 63:259–262.

Jones, R. J., Hoegh-Guldberg, O., Larkum, A. W. D., & Schreiber, U. (1998). Temperature-induced bleaching of corals begins with impairment of the CO_2 fixation mechanism in zooxanthellae. *Plant Cell & Environment* 21:1219–1230.

Keenan, R., Lamb, D., Woldring, O. et al. (1997). Restoration of plant biodiversity beneath tropical tree plantations in northern Australia. *Forest Ecology & Management* 99:117–131.

Kelly, J. (2006). *Cutting the Kenaf*. ABC Tropical North broadcast. 10 May.

Kemfert, C. & Schill, W. P. (2010). *Methane Mitigation*. Cambridge University Press, Cambridge, UK.

Kenk, G. K. (1992). Silviculture of mixed-species stands in Germany. Special publication of the British Ecological Society – FAO report.

Kennedy, E. V., Ordoñez, A. & Diaz-Pulido, G. (2018). Coral bleaching in the southern inshore Great Barrier Reef: A case study from the Keppel Islands. *Marine and Freshwater Research* 60:191–197.

Kenyon, R. A., Loneragan, N. R., Manson, F. J. et al. (2004). Allopatric distribution of juvenile red-legged banana prawns (*Penaeus indicus* H. Milne Edwards, 1837) and juvenile white banana prawns (*Penaeus merguiensis* De Man, 1888) and inferred extensive migration in the Joseph Bonaparte Gulf, north-west Australia. *Journal of Experimental Marine Biology and Ecology* 309:79–108.

Kershaw, A. P., Clark, J. S., Gill, A. M. & D'Costa, D. M. (2002). A history of fire in Australia. In: Bradstock, P. A., Williams, J. E. & Gill, A. M. (Eds.), *Flammable Australia*, Chapter 1. Cambridge University Press, Cambridge, UK.

Keys, S. J. (2003). Aspects of the biology and ecology of the brown tiger prawn *Penaeus esculentus*, relevant to aquaculture. *Aquaculture* 217:325–334.

Kiese, R., Hewett, B., Graham, A. & Butterbach-Bahl, K. (2003). Seasonal variability of N_2O emissions and CH_4 uptake by tropical rainforest soils of Queensland, Australia. *Global Biogeochemical Cycles* 17:1043–1055.

Kimball, B. (2011). Crop yields and CO_2 fertilization. In: Hillel, D. & Rosenzweig, C. (Eds.), *Handbook of Climate Change and Agroecosystems*, pp. 87–108. Imperial College Press, London, UK.

Kindermann, G., Obersteiner, M., Sohngen, B. et al. (2008). Global cost estimates of reducing carbon emissions through avoided deforestation. *Proceedings of the National Academy of Sciences of the United States of America* 105(30): 10302–10307.

Kingston, G. & Norris, C. (2001). The green cane harvesting system – An Australian perspective. In: *Innovative Approaches to Sugarcane Productivity in the New Millennium*, p. 9. Agronomy Workshop Abstracts, Miami, Florida, American Society of Sugar Cane Technologists, Miami, FL, USA.

Klumpp, D. W., Bayne, B. L. & Hawkins, A. J. S. (1992). Nutrition of the giant clam *Tridacna gigas* (L). I. Contribution of filter feeding and photosynthesis to respiration and growth. *Journal of Experimental Marine Biology and Ecology* 155:105–122.

Koci, J. & Nelson, P. N. (2016). Tropical dairy pasture yield and nitrogen cycling: Effect of urea application rate and a nitrification inhibitor, DMPP. *Crop and Pasture Science* 67:766–779.

Kolkovski, S., Curnow, J. & King, J. (2010). Further development towards commercialisation of marine fish larvae feeds – Artemia. Final Report, Fisheries Research and Development Corporation Project No 2004/239, Western Australia, Department of Fisheries.

Kortschak, H. P., Hartt, C. E. & Burr, G. O. (1965). Carbon dioxide fixation in sugarcane leaves. *Plant Physiology* 40:209–213.

Kramer, C. & Gleixner, G. (2007). Soil organic matter in soil depth profiles: Distinct carbon preferences of microbial groups during carbon transformation. *Soil Biology and Biochemistry* 40:425–433.

Krause, G. H., Winter, K., Krause, B. et al. (2010). High temperature tolerance of a tropical tree, *Ficus insipida*: Methodological reassessment and climate change consideration. *Functional Plant Biology* 37:890–900.

Kurihara, M., Magner, T., Hunter, R. A. & McCrabb, G. J. (1999). Methane production and energy partition of cattle in the tropics. *British Journal of Nutrition* 81:227–234.

Kutt, A. S. & Woinarski, J. C. Z. (2007). The effect of grazing and fire on vegetation and the vertebrate assemblage in a tropical savanna woodland in north-eastern Australia. *Journal of Tropical Ecology* 23:95–106.

Laidlow, M., Kitching, R., Goodall, K. et al. (2007). Temporal and spatial variation in an Australian tropical rainforest. *Austral Ecology* 32:10–20.

Lal, R. (2004). Soil carbon sequestration to mitigate climate change. *Geoderma* 123:1–22.

Lal, R. & Follett, R. F. (Eds.). (2009). *Soil Carbon Sequestration and the Greenhouse Effect*, 2nd ed., Vol. 57. SSSA Special Publication.

Lamb, D. (2014). *Large-Scale Forest Restoration*. Taylor & Francis Group, London, UK.

Lamoureux, S. C., Veneklaas, E. J. & Poot, P. (2016). Informing arid-region mine-site restoration through comparative ecophysiology of *Acacia* species under drought. *Journal of Arid Environments* 133:73–84.

Langton, M. (2012). *The Conceit of Wilderness Ideology*. Australian Broadcasting Corporation, Boyer Lectures, 2012.

Lanyon, J. M., Limpus, C. J. & Marsh, H. (1989). Dugongs and turtles: Grazers in the seagrass system. In: Larkum, A. W. D., McComb A. J. & Shepherd, S. A. (Eds.), *Biology of Seagrasses: A Treatise on the Biology of Seagrasses with Special Reference to the Australian Region*, pp. 610–634. Elsevier, Amsterdam, UK and New York, USA.

Lassiere, O. & McCredie, A. (1983). A survey of the vegetation and soils of the Hilton mine area. MIM Research & Development Technical Report. July/August.

Laurance, W. F., Oliveira, A. A., Laurance, S. G. et al. (2004). Pervasive alteration of tree communities in undisturbed Amazonian forests. *Nature* 428:171–175.

Lawn, R. J. & Cottrell, A. (1988). Wild mung bean and its relatives in Australia. *Biologist* 35:267–273.

Lawn, R. J. & Imre, B. C. (1991). Crop improvements for tropical and subtropical Australia: Designing plants for different climates. *Field Crops Research* 26:113–139.

Lawn, R. J. & Russell, J. S. (1978). Mung beans: A grain legume for summer rainfall cropping areas of Australia. *The Journal of the Australian Institute of Agricultural Science* 44:28–41.

Lawn, R. J. & Watkinson, A. R. (2002). Habitats, morphological diversity and distribution of the genus *Vigna* in Australia. *Australian Journal of Agricultural Research* 53:1305.

Lawrence, P. J. (2004). Modelling the climate impact of Australian land cover changes. PhD thesis, University of Queensland, Brisbane, Queensland, Australia.

Lee Long, W. J., McKenzie, L. J., Roelofs, A. J. et al. (1996). *Baseline Survey of Hinchinbrook Region Seagrasses*. Research Publication No. 51, Great Barrier Reef Marine Park Authority, Australia.

Le Quere, C., Rodenbeck, C., Buitenhuis, E. T. et al. (2007). Saturation of the Southern Ocean CO_2 sink due to recent climate change. *Science* 316:1735–1738.

Levas, S., Grottoli, A. G., Warner, M. E. et al. (2015). Organic carbon fluxes mediated by corals at elevated pCO_2 and temperature. *Marine Ecology Progress Series* 519:153–164.

Liebman, M. & Dyck, E. (1993). Crop rotation and intercropping strategies for weed management. *Ecological Applications* 3:92–122.

Liedloff, A. C. & Cook, G. D. (2011). The interaction of fire and rainfall variability on tree structure and carbon fluxes in savannas: Application of the Flames model. In: Hill, M. J. & Hanan, N. P. (Eds.), *Ecosystem Functions in Savannas*, pp. 293–308. CRC/Taylor Francis, Boca Raton, FL, USA.

Limpus, C. J. & Nicholls, N. (1988). The Southern Oscillation regulates the annual numbers of green sea turtles (*Chelonia mydas*) breeding around northern Australia. *Australian Journal of Wildlife Research* 15:157–161.

Lindenmayer, D. B., Franklin, J. F., Löhmus, A. et al. (2012). A major shift to the retention approach for forestry can help resolve some global forest sustainability issues. *Conservation Letters*. doi:10.1111/j.1755-263x2012.00257.x.

Lindsay, A. (2011). Enthusiasm emerging for the African mahogany plantation industry up north. *Australian Forest Grower* 34:25–26.

Lindsay, S. (2018). The conservative case against live animal exports. *Quadrant* LXII:28–32 (June 2018. No. 547).

Long, J., Archer, M., Flannery, T. & Hand, S. (2002). *Prehistoric Mammals of Australia and New Guinea: One Hundred Years of Evolution*. Johns Hopkins University Press, Baltimore, MD, USA.

Lottermoser, B. G. (2011). Colonisation of the rehabilitated Mary Kathleen uranium mine site by *Calotropis procera*: Toxicity risk to grazing animals. *Journal of Geochemical Exploration* 111:39–46.

Lottermoser, B. G. & Ashley, P. M. (2005). Tailings dam seepage at the rehabilitated Mary Kathleen uranium mine, Australia. *Journal of Geochemical Exploration* 85:119–137.

Loxton, E., Schirmer, J. & Kanowski, P. (2012). Employment of Indigenous Australians in the forestry sector: A case study from northern Australia. *Australian Forestry* 75:73–81.

Lucas, J. S. (1982). Quantitative studies of feeding and nutrition during larval development of the coral reef asteroid *Acanthaster planci* (L.). *Journal of Experimental Marine Biology and Ecology* 65:173.

Lucas, J. S. (1994). The biology, exploitation and mariculture of giant clams (Tridacnidae). *Fisheries Science & Aquaculture* 2:181–223.

Lucas, J. S. (2013). Crown-of-thorns starfish. *Current Biology* 23:945–946.

Lugo, A. E., Brown, S. & Chapman, J. (1988). An analytical review of production rates and stem wood biomass of tropical forest plantations. *Forest Ecology & Management* 23:179–200.

Luong-Van, T., Renaud, S. M. & Parry, D. L. (1999). Evaluation of recently isolated Australian tropical microalgae for the enrichment of the dietary value of brine shrimp (*Artemia*) nauplii. *Aquaculture* 170:161–173.

Luthi, D., Le Floch, M., Bereiter, B. et al. (2008). High resolution [CO_2] record 650,000–800,000 years before present. *Nature* 453:379–382.

Löttge, U. (1997). *Physiological Ecology of Tropical Plants*. Springer-Verlag, Berlin, Germany.

Luyssaert, S., Schultzde, E. D., Börner, A. et al. (2008). Old-growth forests as global carbon sinks. *Nature Letters* 455:213–215.

Major, J., Rondon, M., Molina, D. et al. (2010). Maize yield and nutrition during 4 years after biochar application to a Colombian savanna oxisol. *Plant Soil* 333:117–128.

Maltby, J. E. & Barnes, J. E. (1986). The effect of rate and time of application of nitrogen on Starbonnet rice growth and soil nitrogen changes in the lower Burdekin Area, North Queensland. Final Report for the Rural Industries Research Fund, p. 19.

Marlow, D. (2016). Rehabilitation of land disturbed by mining and extractive industries in Queensland: Some needed legislative and management reforms. *Proceedings of the Royal Society of Queensland* 121:39–52.

Marsh, H., Harris, A. N. M. & Lawler, I. R. (1997). The sustainability of the Indigenous dugong fishery in Torres Strait, Australia/Papua New Guinea. *Conservation Biology* 11:1375–1386.

Martin, G. R. & Twigg, L. E. (2002). Sensitivity to sodium fluoroacetate (1080) of native animals from north-western Australia. *Wildlife Research* 29:75–83.

Matthews, E. & Fung, I. (1987). Methane emissions from natural wetlands: Global distribution, area and environmental characteristics of sources. *Global Biogeochemical Cycles* 1:61–86.

McAlpine, C. A., Syktus, J. I., Deo, R. C. et al. (2007). Modelling the impact of historical land cover change on Australia's regional climate. *Geophysical Research Letters* 34. doi:10.1029/2007GLO 31524.

McCrabb, G. J. & Hunter, R. A. (1999). Prediction of methane emissions from beef cattle in tropical production systems. *Australian Journal of Agricultural Research* 50:1335–1339.

McCullough, C. D. & Lund, M. A. (2006). Opportunities for sustainable mining pit lakes in Australia. *Mine Water and the Environment* 25:220–226 (Technical Communication IMWA Springer-Verlag).

McElwain, J. C., Beerling, D. J. & Woodward, F. I. (1999). Fossil plants and global warming at the Triassic-Jurassic boundary. *Science* 285:1386–1390.

McIvor, J. G. (2001). Litterfall from trees in semi-arid woodlands of north-East Queensland. *Austral Ecology* 26:150–155.

McIvor, J. G. (2003). Competition affects survival and growth of buffel grass seedlings – Is buffel grass a colonizer or an invader? *Tropical Grasslands* 37:176–181.

McKenna, S. A., Jarvis, J., Sankey, T. et al. (2015). Declines of seagrass in a tropical harbour, north Queensland, Australia, are not the result of a single event. *Journal of Biosciences* 40:389–398.

McPhee, D. (2008). *Fisheries Management in Australia*. Federation Press, Sydney, Australia.

Meehl, G. A., Stocker, T. F., Collins, W. D. et al. (2007). Global climate projections. In: Solomon, S., Qin, D., Manning, M. et al. (Eds.), *Climate Change 2007. The Physical Science Basis*. Contributions of Working Group I to the Fourth Assessment Report of the Intergovernmental Panel on Climate Change, pp. 748–845. Cambridge University Press, Cambridge, UK.

Metcalfe, D. J. & Bradford, M. G. (2008). Rainforest recovery from dieback, Queensland, Australia. *Forest Ecology and Management* 256:2073–2077.

Mifsud, G. (1999). Ecology of the Julia Creek dunnart *Sminthopsis douglasi* (Marsupialia, Dasyuridae). MSc thesis, La Trobe University, Australia.

Milich, L. (1999). The role of methane emissions in global warming: Where might mitigation strategies be focused? *Global Environmental Change* 9:179–201.

Mitchell, A. L., Lucas, R. M., Donnelly, B. E. et al. (2007). A new map of mangroves for Kakadu National Park, Northern Australia based on stereo aerial photography. *Aquatic Conservation* 17:446–467.

Mitchell, D. S. & Turr, N. M. (1975). The rate of growth of *Salvinia molesta* (*S. auriculate* Act) in laboratory and natural conditions. *Journal of Applied Ecology* 12:213–225.

Mitra, S., Wassmann, R. & Viek, P. L. G. (2005). An appraisal of global wetland area and its organic carbon stock. *Current Science* 88:25–35.

Molenaar, S. (2018). Special report, Australian Horticultural Exporters and Importers Association.

Mooney, S. D., Harrison, S. P., Bartlein, P. J. et al. (2011). Late Quaternary fire regimes in Australia. *Quaternary Science Reviews* 30:28–46.

Mooney, S. D., Harrison, S. P., Bartlein, P. J. et al. (2012). The prehistory of fire in Australasia. In: Bradstock, R. A., Gill, A. M. & Williams, R. J. (Eds.), *Flammable Australia: Regimes, Biodiversity and Ecosystems in a Changing World*, pp. 3–25. CSIRO Publishing, Collingwood, Victoria, Australia.

Morello, E. B., Plagányi, E. E., Babcock, R. C. et al. (2014). Model to manage and reduce crown-of thorns starfish outbreaks. *Marine Ecology Progress Series* 512:167–183.

Morley, R. J. (2000). *Origin and Evolution of Tropical Rain Forests*. John Wiley & Sons, Ltd., Chichester, UK.

Mott, J. J., Williams, J., Andrew, M. H. & Gillison, A. W. (1985). Australian savanna ecosystems. Tothill, J. C, & Mott, J. J. (Eds.), *Ecology and Management of the World's Savannas*, pp. 56–82. Australian Academy of Science, Canberra.

Mucha, S. B. (1979). Estimations of tree ages from growth rings of eucalypts in northern Australia. *Australian Forestry* 42:13–16.

Muchow, R. C. (1981). The growth and culture of kenaf (*Hibiscus cannabinus* L.) in tropical Australia. In: Wood, I. M. & Stewart, G. A. (Eds.), *Kenaf as a Potential Source of Pulp in Australia*, pp. 10–28. Proceedings of Kenaf Conference, CSIRO, Brisbane, Queensland, Australia.

Muchow, R. C. & Coates, D. B. (1986). An analysis of the environmental limitation to yield of irrigated grain sorghum during the dry season in tropical Australia using a radiation interception model. *Australian Journal of Agricultural Research* 37:135–148.

Mudd, G. M. & Patterson, J. (2010). Continuing pollution from the Rum Jungle U-Cu project: A critical evaluation of environmental monitoring and rehabilitation. *Environmental Pollution* 158:1252–1260.

Murphy, A. M. & Colucci, P. E. (1999). A tropical forage solution to poor quality ruminant diets: A review of *Lablab purpureus*. *Livestock Research for Rural Development* 11. http://www.cipar.org.co/lrrdl11/2/colul12htm.

Myers, R. J. K. & Robbins, G. B. (1991). Sustaining productive pastures in the tropics. 5. Maintaining productive sown grass pastures. *Tropical Grasslands* 25:104–110.

Naiman, R. J., Bunn. S. E., Nilsson, C. et al. (2002). Legitimizing fluvial ecosystems as users of water: An overview. *Environmental Management 30*:455–467.

Neale, T. (2017). Re-reading the Wild Rivers Act controversy. In: Vincent, E. & Neale, T. (Eds.), *Unstable Relations: Indigenous People and Environmentalism in contemporary Australia*, pp. 25–53. UWA Publishing, Crawley, Western Australia.

Neilson, E. H., Edwards, A. M., Blomstedt, C. et al. (2015). Utilization of a high-throughput shoot imaging system to examine the dynamic phenotypic responses of a C4 cereal crop plant to nitrogen and water deficiency over time. *Journal of Experimental Botany* 66:1817–1832.

Newton, J. R., Smith-Keune, C. & Jerry, D. R. (2010). Thermal tolerance varies in tropical and sub-tropical populations of barramundi (*Lates calcarifer*) consistent with local adaptation. *Aquaculture* 308:S128–S132.

Nickles, D. G., Spidy, T., Rider, E. J. et al. (1983). Genetic variation and windfirmness among provenances of *Pinus caribeae* Mor. var. *hondurensis* Barr. and Golf. in Queensland. *Silviculture* 29:125–130.

Nightingale, J. M., Hill, M. J., Phinn, S. R. et al. (2008). Use of 3-PG and 3-PGS to simulate forest growth dynamics of tropical rainforests. I. Parameterisation and calibration for old-growth, regenerating and plantation forests. *Forest Ecology and Management* 254:107–121.

Norby, R. J. & Luo, Y. (2004). Evaluating ecosystem responses to rising atmospheric CO_2 and global warming in a multi-factor world. *New Phytologist* 162:281–293.

Norton, J. H., Shepherd, M. A., Long, H. M. & Fitt, W. K. (1992). The zooxanthellae tubular system in the giant clam. *Biological Bulletin* 183:503–506.

Oelkers, E. H. & Cole, D. R. (2008). Carbon dioxide sequestration: A solution to a global problem. *Elements* 4:305–310.

O'Hara, I. M. (2010). The potential for ethanol production from sugar-cane in Australia. Proceedings of the Australian Society of Sugar Cane Technologists, Bundaberg, Queensland, Australia, pp. 600–609.

Orchard, K. A., Cernusak, L. A. & Hutley, L. B. (2010). Photosynthesis and water-use efficiency of seedlings from northern Australian monsoon forests, savanna and swamp habitats grown in a common garden. *Functional Plant Biology* 37:1050–1060.

Palmer, B. & Vogler, W. (2012). *Cryptostegia grandifolia* (Roxb.) R. Br. – rubber vine. CSIRO Publishing. http://www.publish.csiro.au.

Pan, Y. D., Birdsey, R. A., Fang, J. et al. (2011). A large and persistent carbon sink in the world's forests. *Science* 333:988–993.

Paynter, Q. & Flanagan, G. J. (2004). Integrating herbicide and mechanical control treatments with fire and biological control to manage an invasive wetland shrub, *Mimosa pigra*. *Journal of Applied Ecology* 41:615–629.

Pegg, K. G., Moore, N. Y. & Bentley, S. (1996). *Fusarium* wilt of banana in Australia: A review. *Australian Journal of Agricultural Research* 47:637–650.

Peterson, B. J. & Fry, B. (1987). Stable isotopes in ecosystem studies. *Annual Review of Ecology and Systematics* 18:293–320.

Phillips, O. L., Aarago, L. E. O. C., Lewis, S. L. et al. (2009). Drought sensitivity of the Amazon rainforest. *Science* 323:1344–1347.

Poiner, I. R., Buckworth, R. C. & Harris, A. N. M. (1990). Incidental capture and mortality of sea turtles in Australia's Northern Prawn Fishery. *Australian Journal of Marine and Freshwater Research* 41:97–110.

Pollard, P. C. & Greenway, M. (2013). Seagrasses in tropical Australia, productive and abundant for decades, decimated overnight. *Journal of Biosciences* 38:157–166.

Power, I. M., Harrison, A. L., Dipple, G. M. et al. (2013). Carbon mineralisation: From natural analogues to engineered systems. *Reviews in Mineralogy and Geochemistry* 77:305–360.

Preen, A. R. & Marsh, H. (1995). Response of dugongs to large-scale loss of seagrass from Hervey Bay, Queensland, Australia. *Wildlife Research* 22:507–519.

Primack, R. & Corlett, R. (2005). *Tropical Rain Forests: An Ecological and Biogeographical Comparison*. Blackwell Publishing, Oxford, UK.

Prior, L. D., Eamus, D. & Bowman, D. M. J. S. (2004). Tree growth rates in northern Australia savanna habitats: Seasonal patterns and correlations with leaf attributes. *Australian Journal of Botany* 52:303–314.

Puschendorf, R., Carnaval, A. C., Van Der Wal, T. et al. (2011). Disease-driven amphibian extinctions. *Conservation Biology* 25(5):956–964.

Pyne, S. J. (1991). *Burning Bush: A Fire History of Australia*. University of Washington Press, Seattle, WA, USA.

Rädle, M & Wissemeier, L. (2001). 3,4-dimethylpyrazole phosphate (DMPP) – A new nitrification inhibitor for agriculture and horticulture. An introduction. *Biology and Fertility of Soils* 34:79–84.

Radomiljac, A. M. (1998). The influence of pot-host species, seedling age and supplementary nursery nutrition on *Santalum album* Linn. (Indian sandalwood) plantation establishment within the Ord River Irrigation Area, Western Australia. *Forest Ecology and Management* 102:193–201.

Radomiljac, A. M., McComb, J. A. & McGrath, J. F. (1999). Intermediate host influence on the root hemi-parasite *Santalum album* L. biomass partitioning. *Forest Ecology and Management* 113:143–153.

Radunz, B. L., Wilson, G. & Beere, G. (1984). Feeding rubberbush (*Calotropis procera*) to cattle and sheep (Northern Australia). *Australian Veterinary Journal* 61:243–244.

Rai, S. N. (1990). Status and cultivation of sandalwood in India. In: Hamilton, L. & Conrad, C. E. (Eds.), *Proceedings of the Symposium on Sandalwood in the Pacific*. USDA Forest Service General Technical Paper PSW-122, Honolulu, HI, USA, pp. 66–71.

Redden, R. J., Kroonenberg, P. M. & Basford, K. E. (2012). Adaptation analysis of diversity in adsuki germplasm introduced into Australia. *Crop & Pasture Science* 63:142–154.

Réjou-Méchain, M., Tanguy, A., Chave, C. & Hérault, B. (2017). BIOMASS: An R package for estimating above-ground biomass and its uncertainty in tropical forests. doi:10.1111/2041-210X.12753.

Richards, A. E., Andersen, A. N., Schatz, R. et al. (2012). Savanna burning, greenhouse gas emissions and indigenous livelihoods, including the Tiwi Carbon Study. *Austral Ecology* 37:712–723.

Richards, A. E., Cook, G. D. & Lynch, B. T. (2011). Optimal fire regimes for soil carbon storage in tropical savannas of northern Australia. *Ecosystems* 14:503–518.

Rimmer, M. (2003). Barramundi. In: Lucas, J. S. & Southgate, P. C. (Eds.), *Aquaculture: Farming Aquatic Animals and Plants*, Chapter 18. Fishing News Books, Blackwell Publishing Ltd., Oxford, UK.

Rindos, D. (1980). Symbiosis, instability and the origins and spread of agriculture: A new model. *Current Anthropology*, 21:751–772.

Roberts, J. (2008). *Massacres to Mining: The Colonisation of Aboriginal Australia*. Impact Investigative Media Productions, Bristol, UK.

Robertson, A. I. & Blaber, S. J. M. (1992). Plankton, epibenthos and fish communities. In: Robertson, A. I. & Alongi, D. M. (Eds.), *Tropical Mangrove Ecosystems* (Coastal and Estuarine Studies), pp. 173–224. American Geophysical Union, Washington DC, USA.

Robertson, A. I. & Daniel, P. A. (1989). The influence of crabs on litter processing in high intertidal mangrove forests in tropical Australia. *Oecologia* 78:191–198.

Robertson, F. (2003). Sugarcane trash management: Consequences for soil carbon and nitrogen. Report to CRC for Sustainable Sugar Production, Townsville, Queensland, Australia.

Robins, J. B., Halliday, I. A., Staunton-Smith, J. et al. (2005). Freshwater flow requirements of estuarine fisheries in tropical Australia: A review of the state of knowledge and application of a suggested approach. *Marine and Freshwater Research* 56:343–360.

Robins, L. & Kanowski, P. (2011). "Crying for our Country": Eight ways in which "Caring for our Country" has undermined Australia's regional model for natural resource management. *Australasian Journal of Environmental Management* 18:88–108.

Robinson, J., Popova, E. E., Yool, A. et al. (2014). How deep is deep enough? Ocean iron fertilisation and carbon sequestration in the Southern Ocean. *Geophysical Research Letters* 41:2489–2495.

Room, P. M., Forno, I. W. & Taylor, M. F. J. (1984). Establishment in Australia of two insects for biological control of the floating weed *Salvinia molesta*. *Bulletin of Entomological Research* 74:505–516.

Room, P. M., Harley, K. L. S., Forno, I. W. & Sands, D. P. A. (1981). Successful biological control of the floating weed Salvinia. *Nature* 294:78–80.

Rosentreter, J., Maher, D. T., Erler, D. V. et al. (2018). Methane emissions partially offset "blue carbon" burial in mangroves. *Science Advances* 4:eaao4985.

Ross, H. (1992). Opportunities for aboriginal participation in Australian social impact assessment. 10.1080/07 349165. 1992. 9725731

Rowe, C. (2015). Late Holocene swamp transition in the Torres Strait, northern tropical Australia. *Quaternary International* 385:56–68.

Ruschena, L. J., Stacey, G. S., Hunter, G. D. & Whitman. P. C. (1974). Research into revegetation of concentrator tailings dams at Mount Isa. Presentation at AIMM Regional Meeting, August.

Russell, D. J. & Garrett, R. N. (1985). Early life history of barramundi, *Lates calcarifer* (Bloch) in north-eastern Queensland. *Australian Journal of Marine and Freshwater Research* 36:191–201.

Russell, J. S. (1976). Use of pattern analysis in the selection of crop homoclimates. *Proceedings of Grains Research Conference* 2a(iii):1–3.

Russell-Smith, J., Murphy, B. P., Meyer, C. P. et al. (2009). Improving estimates of savanna burning emissions for greenhouse accounting in northern Australia: Limitations, challenges, applications. *International Journal of Wildland Fire* 18:1–18.

Russell-Smith, J., Yates, C., Edwards, A. et al. (2003). Contemporary fire regimes of northern Australia, 1997–2001: Change since Aboriginal occupancy, challenges for sustainable management. *International Journal of Wildland Fire* 12:283–297.

Ryan, T. & Shea, G. (1977). Exotic pine plantations in Queensland and the role of the Beerwah-Beerburram Forests. Post consultation tour notes No. 8. Third World Consultation of Forest Tree Breeding, Canberra.

Sakalidis, M. L., Ray, J. D., Lanoiselet, V. et al. (2011). Pathogenic Botryosphaeriaceae associated with *Mangifera indica* in the Kimberley region of Western Australia. *European Journal of Plant Pathology* 130:379–391.

Salisbury, F. B. & Ross, C. W. (1992). Chapter 14. Assimilation of nitrogen and sulphur. In: Salisbury, F. B. & Ross, C. W. (Eds.), *Plant Physiology*, 4th ed., pp. 289–307. Wadsworth, Inc., Belmont, CA, USA.

Sanger, D. D. & Kirkpatrick, J. B. (2016). Moss and vascular epiphyte distributions over host tree and elevation gradients in Australian sub-tropical rainforests. *Australian Journal of Botany* 63:696–704.

Sattler, P. & Williams, R. (Eds.). (1999). *The Conservation Status of Queensland's Bioregional Ecosystems*. Environmental Protection Agency, Queensland Government, Brisbane, Queensland, Australia.

Sattler, P. S. (2014). *Five Million Hectares – A Conservation Memoir 1972–2008*. Paul Sattler Eco-consulting Pty. Ltd., Mount Cotton, Queensland, Australia.

Scherer-Lorenzen, M., Potvin, C., Koricheva, J. et al. (2005). The design of experimental tree plantations for functional biodiversity research. doi:10.1007/3-540-26599-6_16.

Scherm, H. & Coakley, S. M. (2003). Plant pathogens in a changing world. *Australasian Plant Pathology* 32:157–165.

Schippers, P., Vlam, M., Zuidema, P. A. & Sterck, F. (2015). Sapwood allocation in tropical trees: A test of hypotheses. *Functional Plant Biology* 42:697–709.

Schlenker, W. & Roberts, M. J. (2009). Non-linear temperature effects indicate severe damage to U.S. crop yields under climate change. *Proceedings of the National Academy of Sciences of the United States of America* 106:15594–15598.

Schlesinger, W. H. (1990). Evidence from chronosequence studies for a low carbon storage potential of soils. *Nature* 348:232–234.

Schwenke, G. D., Ayre, L., Mulligan, D. R. & Bell, L. C. (2000). Soil stripping and replacement for the rehabilitation of bauxite-mined land at Weipa. II. Soil organic matter dynamics in mine soil chronosequences. *Australian Journal of Soil Research* 38:371–394.

Schwenke, G. D., Mulligan, D. R. & Bell, L. C. (1999). Soil stripping and replacement for the rehabilitation of bauxite-mined land at Weipa. I. Initial changes to soil organic matter and related parameters. *Australian Journal of Soil Research* 38:345–369.

Shaw, N. H. (1961). Increased beef production from Townsville lucerne (*Stylosanthes sundaica*, Taub) in the spear grass pastures of central coastal Queensland. *Australian Journal of Experimental Agriculture and Animal Husbandry* 1:73–80.

Shelton, M. & Dalzell, S. (2007). Production, economic and environmental benefits of leucaena pastures. *Tropical Grasslands* 41:174–190.

Sherwin, G. L., George, L., Kannangara, K. et al. (2012). Impact of industrial-age climate change on the relationship between water uptake and tissue nitrogen in eucalypt seedlings. *Functional Plant Biology* 40:201–212.

Shinozaki, K., Yoda, K., Hozumi, K. & Kira, T. (1964). A quantitative analysis of plant form – The pipe model theory. II. Further evidence of the theory and its application in forest ecology. *Japanese Journal of Ecology* 14:133–139.

Simmonds, N. W. & Shepherd, K. (1955). The taxonomy and origin of the cultivated banana. *The Journal of the Linnean Society (Botany)* 55:302–312.

Sjögerstein, S., Black, C. R., Evers, S. et al. (2014). Tropical wetlands: A missing link in the global carbon cycle. *Global Biochemical Cycles* 28:1371–1386.

Slik, J. W. F., Paoli, G., McGuire, K. et al. (2013). Large trees drive forest above-ground biomass variation in moist lowland forests across the tropics. *Global Ecology and Biogeography* 22:1261–1271.

Slot, M., Garcia, M. N. & Winter, K. (2016). Temperature responses of CO_2 exchange in three tropical tree species. *Functional Plant Biology* 43:468–478.

Smillie, R. M. & Nott, R. (1979). Heat injury in leaves of alpine, temperate and tropical plants. *Australian Journal of Plant Physiology* 6:135–141.

Smith, A. (1994). *This El Dorado of Australia*. Studies in North Queensland History No. 20. Department of History and Politics, James Cook University of North Queensland.

Sohngen, B. & Brown, S. (2008). Extending timber rotations: Carbon and cost implications. *Climate Policy* 8:435–451.

Staples, L., Smith, M. & Pontin, K. (2003). Use of zinc sulphide to overcome rodent infestations. In: Wright, E. J., Webb, M. C. & Highlry, E. (Eds.), *Stored Grains in Australia 2003: Proceedings of the Australian Postharvest Technical Conference, Canberra, 25–27 June 2003*. CSIRO Stored Grain Research Laboratory, Canberra, Australia.

Steffen, W., Burbidge, A. A., Hughes, L. et al. (2009). Australia's biodiversity and climate change: a strategic assessment of the vulnerability of Australia's biodiversity to climate change. A report to the National Resource Management Ministerial Council commissioned by the Australian Government. CSIRO Publishing, Collingwood, Victoria, Australia.

Stork, N. E. (2007). Australian tropical forest canopy crane: New tools for new frontiers. *Austral Ecology* 32:4–9.

Stork, N. E., Balston, J., Farquhar, G. D. et al. (2007). Tropical rainforest canopies and climate change. *Austral Ecology* 32:105–112.

Streamer, M., Griffiths, D. J. & Luong-Van, T. (1988). The products of photosynthesis by zooxanthellae (*Symbiodinium microadriaticum*) of *Tridacna gigas* and their transfer to the host. *Symbiosis* 6:237–252.

Strickland, G. R. & Annells, A. J. (1999). Transgenic cotton research paves the way for a new industry in the Kimberley. *Journal of the Department of Agriculture, Western Australia*, Series 4, 40:article 3.

Subedi, R., Akbar, D., Ashwath, N. et al. (2017). Assessing the viability of growing *Agave tequilana* as a biofuel feedstock in Queensland, Australia. *International Journal of Energy Economics and Policy* 7:172–180.

Sun, D. & Dickinson, G. R. (1995). Salinity effects on tree growth, root distribution and transpiration of *Casuarina cunninghamiana* and *Eucalyptus camaldulensis* planted on a saline site in tropical north Australia. *Forest Ecology and Management* 77:127–138.

Syktus, J. I. & McAlpine, C. A. (2016). More than carbon sequestration: Biophysical climate benefits of restored savanna woodlands. *Scientific Reports* 6:29194.

Szabo, S. & Smyth, D. (2003). Indigenous protected areas in Australia: Incorporating Indigenous owned land into Australia's national system of protected areas. Fifth World Parks Congress: Sustainable Finance Stream, September, Durban, South Africa.

Taylor, G., Spain, A., Nefiodoras, A. et al. (2003). Determination of the reasons for deterioration of the Rum Jungle waste rock cover. Report for the Australian Centre for Mining Environmental Research, Kenmore, Queensland, Australia.

Thiagalingam, K., Dalgliesh, N. P., Gould, N. S. et al. (1996). Comparison of no-tillage and conventional tillage in the development of sustainable farming systems in the semi-arid tropics. *Australian Journal of Experimental Agriculture* 36:995–1002.

Thomas, C. D., Cameron, A., Green, R. E. et al. (2004). Extinction risk from climate change. *Nature* 427:145–148.

Thomas, L., Kendrick, G. A., Stat, M. et al. (2014). Population genetic structure of the *Pocillopora damicornis* morphospecies along Ningaloo Reef, Western Australia. *Marine Ecology Progress Series* 513:111–119.

Tielbörger, K. & Salguero-Gómez, R. (2014). Some like it hot: Are desert plants indifferent to climate change? Lüttge, U. Beyschlag, W. & Cushman, J. (Eds.), *Progress in Botany* vol. 75, 377–400. Springer Nature, Heidelberg, Germany.

Tingey-Holyoak, J., Burritt, R. L. & Pisaniello, J. D. (2013). Living with surface water shortage and surplus: The case for sustainable agricultural water storage. *Australasian Journal of Environmental Management* 20:208–224.

Tingley, R., Phillips, B., Letnie, M. et al. (2013). Identifying optimal barriers to halt the invasion of cane toads (*Rhinella marina*) in arid Australia. *Journal of Applied Ecology* 50:129–137.

Tissue, D. T., Thomas, R. B. & Strain, B. R. (1993). Long-term effects of elevated CO_2 and nutrients on photosynthesis and rubisco in loblolly pine seedlings. *Plant Cell & Environment* 16:859–865.

Tothill, J. C. (1971). A review of fire in the management of native pasture with particular reference to north-eastern Australia. *Tropical Grasslands* 5:1–10.

Townsend, S., Schult, J., Douglas, M. & Lautenschlager, A. (2017). Recovery of benthic primary producers from flow disturbance and its implications for an altered flow regime in a tropical savannah river (Australia). *Aquatic Botany* 136:9–20.

Tracey, J. G. (1982). *The Vegetation of the Humid Tropical Region of North Queensland.* CSIRO Publishing, Melbourne, Australia.

Tracey, J. G. & Webb, L. J. (1975). *Vegetation of the Humid Tropical Region of North Queensland.* CSIRO Australia, Long Pocket Labs, Indooroopilly [15 Maps at 1: 100,000 scale plus key].

Tran, D. B. & Dargusch, P. (2016). *Melaleuca* forests in Australia have globally significant carbon stocks. *Forest Ecology and Management* 375:239–237.

Trebeck, K. A. (2007). Tools for the disempowered? Indigenous leverage over mining companies. *Australian Journal of Political Science* 42:541–562.

Trewin, R. (2014). Australian-Indonesian live cattle trade – What future? *Asia and the Pacific Policy Studies* 1:423–430.

Turner, J. & Lambert, M. (2016). Changes in Australian forestry and forest products research for 1985–2013. *Australian Forestry* 79:53–58.

Turton, S., Dickson, T., Hadwen, W. et al. (2010). Developing an approach for tourism climate change assessment: Evidence from four contrasting Australian case studies. *Journal of Sustainable Tourism* 18:429–447.

Twigg, L. E., Martin, G. R., Wilson, N. et al. (2001). Longevity of zinc sulphide on wheat bait in tropical Australia. *Wildlife Research* 28:261–267.

Twilley, R. R., Dhen, R. H. & Hargis, T. (1992). Carbon sinks in mangroves and their implications to carbon budget of tropical coastal ecosystems. In: Wisniewski, J. & Lugo, A. E. (Eds.), *Natural Sinks of CO_2*. Workshop: Palmas Del Mar, Puerto Rico, 24–27 February. Reprinted from Water, *Air and Soil Pollution* 64(1–2). Kluwer Academic Publishers, Dordrecht, The Netherlands.

Unger, C. & Laurencourt, T. (2003). Development of a sustainable rehabilitation strategy for the management of acid rock drainage at the historic Mount Morgan gold and copper mine, Central Queensland. *Proceedings of the 6th International Conference on Acid Rock Drainage* pp. 685–692. Cairns, Queensland, Australia, 14–17 July.

Unwin, G. L. & Kriedemann, P. E. (1990). Drought tolerance and rainforest tree growth on a north Queensland rainfall gradient. *Forest Ecology & Management* 30:113–123.

Van Oosterzee, P., Dale, A. & Preece, N. D. (2014). Integrating agriculture and climate change mitigation at landscape scale: Implications from an Australian case study. *Global Environmental Change* 29:306–317.

Van Vreeswyk, A. M. E., Leighton, K. A., Payne, A. L. & Hennig, P. (2004). An inventory and condition survey of the Pilbara region. Technical Bulletin, Department of Agriculture and Food, Government of Western Australia.

Veron, J. E. N., Heogh-Guldberg, O., Lenton, T. M. et al. (2009). The coral reef crisis: The critical importance of <350 ppm CO_2. *Marine Pollution Bulletin* 58:1428–1436.

Vickers, H., Gillespie, M. & Gravina, A. (2012). Assessing the development of rehabilitated grasslands on post-mined landforms in north-west Queensland, Australia. *Agriculture, Ecosystems and Environment* 163:72–84.

Vitelli, J. S. (1992). Trial site layout and suggested strategies for rubber vine control. In: The Control and Management of Rubber Vine – Proceedings of "Larkspur Field Day", Dalrymple Landcare Committee, May, 1992.

Vitelli, J. S. (1995). Rubber vine. In: March, N. (Ed.), *Exotic Weeds and Their Control in North-West Queensland*, pp. 14–17. Queensland Department of Lands, Mount Isa, Queensland, Australia.

W.A. Government. (2017). Buffel grass pastures in the Kimberley. https://www.agric.wa.gov.au.

Walker, G. R. & Mallants, D. (2014). Methodologies for investigating gas in water bores and links to coal seam gas development. CSIRO Land and Water Flagship Report, CSIRO, Australia.

Wang, Z., Chappellaz, J., Park, K. & Mak, J. E. (2010). Large variations in Southern Hemisphere biomass burning during the last 650 years. *Science* 330:1663–1666.

Ward-Fear, G., Brown, G. P., Pearson, D. J. & Shine, R. (2017). An invasive tree facilitates the persistence of native rodents on an over-grazed floodplain in tropical Australia. *Austral Ecology* 42:385–393.

Ward-Fear, G., Pearson, D. J., Brown, G. P. et al. (2016). Ecological immunisation: In situ training of free-ranging predatory lizards reduces their vulnerability to invasive toxic prey. *Biology Letters* 12:5.

Warfe, D. M., Petit, N. E., Davies, P. M. et al. (2011). The "wet-dry" in the wet-dry tropics drives river ecosystem structure and processes in northern Australia. *Freshwater Biology* 56:2169–2195.

Webb, L. J. (1959). A physiognomic classification of Australian rainforests. *Journal of Ecology* 47:551–570.

Webber, B. L. & Woodrow, I. E. (2004). Cassowary frugivory, seed defleshing and fruit-fly infestation influence the transition from seed to seedling in the rare Australian rainforest tree, *Ryparosa* sp. nov. 1 (Achariaceae). *Functional Plant Biology* 31:505–516.

West, J. M. & Salm, R. V. (2003). Resistance and resilience to coral bleaching: Implications for coral reef conservation and management. *Conservation Biology* 17:956–967.

Western Australia Department of Conservation and Land Management. (1995). Purnululu National Park: Management Plan 1995–2005.

Western Australian Government, Ministry of Primary Industry. (1995). Ord River Irrigation Project: A review of its expansion potential.

Whitbread, A. M., Pengelly, B. C. & Smith, B. R. (2005). An evaluation of three tropical ley legumes and their effect on cereal production and soil nitrogen on clay soils in Queensland, Australia. *Tropical Grasslands* 39:9–21.

White, K., Lucas, R., Hardie, R., Merritt, J. & Kirsch, B. (2013). Challenges and opportunities for constructed stream diversions in the Bowen Basin, Central Queensland. In: *Water and Mining*. The Australasian Institute of Mining and Metallurgy. Publication Series 12/2013.

Whitehead, P. J., Bowman, D. M. J. S., Preece, N., Fraser, F. & Cooke, P. (2003). Customary use of fire by indigenous peoples in northern Australia: Its contemporary role in savanna management. *International Journal of Wildland Fire* 12:415–425.

Whitmore, T. C. & Silva, J. N. M. (1990). Brazilian rain forest timbers are mostly very dense. *Commonwealth Forestry Review* 69:87–90.

Williams, D. McB. (1986). Temporal variation in the structure of reef slope fish communities (central Barrier Reef): Short-term effects of *Acanthaster planci*. *Marine Ecology Progress Series* 28:157–164.

Williams, J. E. & Gill, A. M. (1995). The impact of fire regimes on native forests in eastern New South Wales. Environmental Heritage Series No. 2. NSW Parks & Wildlife Service, Hurstville.

Williams, R., Hutley, L., Cook, G. et al. (2004). Assessing the carbon sequestration of mesic savannas in the Northern Territory, Australia: Approaches, uncertainties and potential impacts of fire. *Functional Plant Biology* 31:415–422.

Williams, R. J., Cook, G. D., Gill, A. M. & Moore, D. H. R. (1999). Fire regime, fire intensity and tree survival in a tropical savanna in northern Australia. *Australian Journal of Ecology* 24:50–59.

Williams, S. E., Bolitho, E. E. & Fox, S. (2003). Climate change in Australian tropical forests: An impending environmental catastrophe. *Proceedings of the Royal Society B: Biological Sciences* 270:1887–1892.

Willmott, W. (2009). *Rocks and Landscapes of the National Parks of North Queensland.* Geological Society of Australia, Queensland Division, Brisbane, Queensland, Australia.

Willmott, W. (2017a). Underground water: Lifeblood of the west. In: Willmott, W., Cook, A. & Neville, B. (Eds.), *Rocks, Landscapes & Resources of the Great Artesian Basin A Handbook for Travellers*. Geological Society of Australia, Queensland Division.

Willmott, W. (2017b). Coal-seam gas: A controversial new industry. In: Willmott, W., Cook, A. & Neville, B. (Eds.), *Rocks, Landscapes & Resources of the Great Artesian Basin: A Handbook for Travellers*. Geological Society of Australia, Queensland Division.

Wilson, S. G., Taylor, J. G. & Pearce, A. F. (2001). The seasonal aggregation of whale sharks at Ningaloo Reef, Western Australia: Currents, migrations and the El Niño/Southern Oscillation. *Environmental Biology of Fishes* 61:1–11.

Winter, K. (1979). Effect of different CO_2 regimes on the induction of Crassulacean acid metabolism in *Mesembryanthemum crystallinum* L. *Australian Journal of Plant Physiology* 6:589–594.

Winter, K. & Holtum, J. A. M. (2015). Cryptic crassulacean acid metabolism (CAM) in *Jatropha curcas*. *Functional Plant Biology* 42:711–717.

Winter, K. & Virgo, A. (1998). Elevated CO_2 enhances growth in the rainforest understorey plant *Piper cordulatum* at extreme low light intensities. *Flora* 193:323.

Winter, K., Wallace, B. J., Stocker, G. C. & Roksandic, Z. (1983). Crassulacean acid metabolism in Australian vascular epiphytes and some related species. *Oecologia* 57:129–141.

Woinarski, J. C. Z. (Ed.). (1990). A survey of the wildlife and vegetation of the Purnululu (Bungle) National Park and adjacent area. Report to the WA Department of Conservation and Land Management.

Woinarski, J. C. Z., Legge, S. L., Fitzsimons, J. A. et al. (2011). The disappearing mammal fauna of northern Australia: Context, cause and response. *Conservation Letters* 4:192–201.

Wood, J. G. (1950). Vegetation. In: Noble, N. S. (Ed.), *The Australian Environment*, 2nd ed. CSIRO, Melbourne, Australia.

Wood, T. E., Cavaleri, M. A. & Reed, S. C. (2012). Tropical forest carbon balance in a warmer world: A critical review spanning microbial to ecosystem scale processes. *Biological Reviews* 87:912–927.

Woodroffe, C. D., Chappell, J., Thom, B. G. & Wolanski, E. (1989). Depositional model of a macrotidal estuary and floodplain, South Alligator River, northern Australia. *Sedimentology* 36:737–756.

Woodrow, I. E. (1994). Optimal acclimation of the C_3 photosynthetic system under enhanced CO_2. *Photosynthesis Research* 39:401.

Woolley, P. A. (2008). Julia Creek dunnart. In: Van Dyck, S. & Strahan, R. (Eds.), *The Mammals of Australia*, pp. 136–137. New Holland Publishers, Frenchs Forest, New South Wales, Australia.

Worboys, S. J. (2006). Rainforest dieback mapping and assessment. 2004 Monitoring Report Including an Assessment of Dieback in High Altitude Rainforests, Rainforest CRC, Cairns, Queensland, Australia.

Worboys, S. J. & Jackes, B. R. (2005). Pollination processes in *Idiospermum australiense* (Calycanthaceae), an arborescent basal angiosperm of Australia's tropical rain forests. *Plant Systematics and Evolution* 251:107–117.

World Nuclear Association Publication. (September 2009). Environmental aspects of uranium mining.

Wrigley, T. J., Farrell, P. D. & Griffiths, D. J. (1991). Ecologically sustainable water clarification at the clear water lagoon, Mount Isa. *Water* 18:32–34.

Xstrata (2010). Mount Isa Mines Sustainability Report for 2010.

Yeates, S. J., Strickland, G. R. & Grundy, P. R. (2013). Can sustainable cotton production be developed for tropical northern Australia? *Crop & Pasture Science* 64:1127–1140.

Yeo, W. L. J. & Fensham, R. J. (2014). Will Acacia secondary forest become rain forest in the Australian Wet Tropics. *Forest Ecology & Management* 331:208–217.

Yibarbuk, D., Whitehead, P. J., Russell-Smith, J. et al. (2001). Fire ecology and Aboriginal land management in central Arnhem Land, northern Australia: A tradition of ecosystem management. *Journal of Biogeography* 28:325–343.

Zann, L. P. & Weaver, K. (1988). An evaluation of control programs on the crown-of-thorns starfish undertaken on the Great Barrier Reef. *Proceedings of the 6th International Coral Reef Symposium,* Vol. 2, p. 183.

Zeng, N. (2008). Carbon sequestration via wood burial. *Carbon Balance & Management* 3(1):1–12.

Ziska, L. H. & Teasdale, J. R. (2000). Sustained growth and increased tolerance to glyphosate observed in a C_3 perennial weed, quackgrass (*Elytrigia repens*) grown at elevated CO_2. *Australian Journal of Plant Physiology* 27:159–166.

Index

A

ACIAR, *see* Australian Centre for International Agricultural Research
Agave, 58–59
Angleton grass, 73
Aquaculture
  barramundi, 82
  giant clams, 83–84
  groundwater management, 82
  microalgal species, 81
  prawns, 83
Artificial lakes, 28–29
Atherton Tablelands, 18, 64
Australian Centre for International Agricultural Research (ACIAR), 66, 83, 121

B

Bauxite mining, 98–99
Beef cattle industry
  Angleton grass, 73
  buffel grass, 71
  CH_4 emission, 72, 73
  live cattle export, 71, 72
  Rhodes grass, 73
  success, 71
Beef production, 69, 70, 71
Billabongs, 23, 26–27, 33, 112, 115
Burdekin River Irrigation Area, 28

C

CAM, *see* Crassulacean acid metabolism
Canopy Crane Research Facility, 4
Cape Tribulation, 4
Cape York Peninsula Land Use Strategy (CYPLUS), 106
Carbon capture and storage, 19
Carbon forestry, 20, 42
Carbon sequestration, 21
Coal-seam gas (CSG), 27, 101
Community Rainforest Restoration Program (CRRP), 63, 64
Compromised forest, 3
Conservation reserves
  joint management, 106
  land management, 107–108
  wilderness, 105
Coronation Hill, 93–94
COTS, *see* Crown-of-thorns

C-4 Photosynthesis, 49–50
Crassulacean acid metabolism (CAM), 7–8, 40, 59
Crops
  animal production, 47
  broadacre
    cotton, 52–53
    rice, 53–54
    sorghum, 54
  climate change, 47
  forest diversity, 66
  fruits and vegetables
    asparagus, 62
    banana, 60–61
    mango, 61
    protected cropping, 60
  legumes (*see* Leguminous plants)
  novel
    agave, 58–59
    kenaf, 58
    sandalwood, 57
  reduce losses, 67
  rotation, 56–57
  sugar cane (*see* Sugar cane industry)
Crop improvements, 47, 48
Crown-of-thorns (COTS), 87
CRRP, *see* Community Rainforest Restoration Program
CSG, *see* Coal-seam gas
Cyclone Larry, 13
CYPLUS, *see* Cape York Peninsula Land Use Strategy

D

Daintree National Park, 5
Daintree Rainforest Observatory, 12, 21
Desert ecosystems, 45–46
Domiculture, 115–116
Drought tolerance, 10, 12

E

Ecological limits of hydrological alterations (ELOHA), 111
ELOHA, *see* Ecological limits of hydrological alterations
Epiphytic ferns, 6, 7
ESCAS, *see* Exporter Supply Chain Assurance Scheme
Exporter Supply Chain Assurance Scheme (ESCAS), 72

F

FACE, *see* Free-air CO$_2$ enrichment
Farm dams, 26–27
Ferns and orchids, 6
Fisheries industry
   barramundi, 80
   indigenous fishing rights, 81
   mud crabs, 80
   Prawn fishery, 80
   recreational fishing, 81
   types, 79
Fisheries Research and Development
        Corporation (FRDC), 78
Flood
   recovery programs/cost, 124
   stock loss, 123
Forest dieback, 22
FRDC, *see* Fisheries Research and
        Development Corporation
Free-air CO$_2$ enrichment (FACE), 16

G

Gall wasp, 67
GBR, *see* Great Barrier Reef
GBR, tourism, 88–89
GCTB, *see* Green Cane Trash Blanketing
Globalisation, 120
Grassland habitats, 55
Great Artesian Basin, 24
Great Barrier Reef (GBR), 29, 77, 78, 79, 80,
        81, 84, 102, 116
Green Cane Trash Blanketing
        (GCTB), 48–49
Groundwater management, 82

H

Herbert River systems, 75
Hydrographic data, 77

I

Indigenous land management, 113–114
Integrated pest management (IPM), 37, 52, 67
Intergovernmental Panel on Climate Change
        (IPCC), 45, 119
Intermittent rivers, 24–25
IPCC, *see* Intergovernmental Panel on Climate
        Change
IPM, *see* Integrated pest management

J

Joint management, 106

K

Kakadu Conservation Zone (KCZ), 93
KCZ, *see* Kakadu Conservation Zone
Kenaf, 58
Kidston gold, 99–100
Kimberley national parks, 107–108

L

Lake Argyle, 28
Lake Julius, 28
Lake Moondarra, 28
Land management, 107–108
Leguminous plants
   crop rotation, 56–57
   fermentable feedstock, 55
   forage, 56
   grassland habitats, 55
Leichhardt River, 28
Litterfall, 21, 69
Liquid natural gas (LNG), 101
Live weight gain (LWG), 73
LNG, *see* Liquid natural gas
LWG, *see* Live weight gain

M

Management
   agro-ecosystem, 111
   government agencies, 111
   land, indigenous, 113–114
   natural, 120
   river flows, 112
   rural industries, 112–113
Mangroves
   formation, 30
   location, 29
   pollen analysis, 30
   rehabilitation, 30
   stratigraphical records, 29
   survival rates, 30–31
Marine
   coral reef, 78
   GBR, 88–89
   GBRMPA, 84
   Ningaloo park, 87–88
   offshore, 117
   resources, 81
   species, 30, 78, 80
MAT, *see* Mean annual temperature
Mean annual temperature (MAT), 13
Melaleuca swamps, 39
Mesotrophic forest, 39
Microalgal species, 81
Mine-site rehabilitation
   aims, 91

history, 100
  legislative aspects, 102–103
Mining
  bauxite, 98–99
  Coronation Hill, 93–94
  Kidston gold, 99–100
  Mount Isa, 95–98
  open-cut methods, 100–101
  Pilbara, 94
  Ranger Uranium Mine, 90–93
  Rum Jungle, 94–95
Mud crabs, 80

N

National parks
  fire scars, 109
  hot spots, 109
  joint management, 109–110
  Kimberley, 107–108
  northern territory, 109
  Pilbara, 108–109
  Queensland, 105–106
National resource management (NRM), 120
Neogene expansion, 1
Net primary production (NPP), 13, 31, 41
NPP, *see* Net primary production
NRM, *see* National resource management
Nutrient cycling, 69

O

Oceanic carbon sequestration, 77–78, 119
Ord River Irrigation Area (ORIA) scheme, 1,
        25, 28, 36, 47, 48, 52, 53, 54, 57, 58,
        61, 122

P

PAR, *see* Photosynthetically active radiation
Pastures
  abandoned, 74–75
  buffel-dominated, 71–72
  crops/legumes, 70
  dairy, 70–71
  grazing, 69–70
  improvement, 69–71
  lablab, 70
Patch death, 21
Photosynthetically active radiation (PAR), 3, 54
Photosynthetic biomass, 40, 118
Pilbara mine, 94
Pilbara Port Authority, 30
Pipe model, 118
Plantation forestry
  cattle grazing, 62
  CRRP, 64

growth rates, 65
selective logging, 63
silviculture, 65–66
timber industry, 62
tropical north, 63–64
understorey component, 64
Pollen analysis, 30
Post-federation developments, 117
Prawn fishery, 80

R

RAC, *see* Resource Assessment Commission
Rainforest
  annual rainfall isohyets, 1
  autochthonous flora, 1
  CAM mechanism, 7–8
  climate change
    annual rainfall, 10
    classification, 8–9
    coastal regions, 10, 11
    cyclones, 13
    leaf phenology, 11–12
    structural forest types, 9–10
    tree mortality., 12
    vegetation, 9
  economic value, 2
  floral biodiversity
    climatic events, 4
    crane plot, 4
    fossil record, 4–5
    lowland plots, 4
    mid-elevation plots, 4
    niche, 5
    observations, 5–6
    taxonomic groups, 5
    vegetation, 4
  geological evidence, 2
  neotropics, 1
  ORIA, 1
  photosynthetic productivity, CO_2
      concentration
    FACE, 16
    fertilising effect, 16
    photon flux densities, 15
    stomata respond, 17
  plant families, 2
  sinks of carbon
    biomass, 19
    deforestation, 18
    forest litter, 21
    human disturbance, 19
    key processes, 21
    phenological groups, 20–21
    soil organic matter, 20
  temperatures effects
    heat Injury, plants, 14

MAT, 13
  NPP, 13
  photosynthetic productivity, 14–15
tree health, 21–22
vascular epiphytes
  dry periods, 6–7
  ferns and orchids, 6
  microclimate factors, 6
  VPD, 6
water-use efficiency, 17–18
WTWHA, 1
Ramsar Convention, 23
Rangelands, 73–74
Ranger Uranium Mine, 90–93
Recreational fishing, 81
REDD, *see* Reduced emissions
          from deforestation and
          forest degradation
Reduced emissions from deforestation and forest
          degradation (REDD), 66, 67, 118
Reef ecosystems
  carbon flux, 85–86
  coral bleaching, 84–85
  crown-of-thorns starfish, 86–87
  giant clams, 86
  Ningaloo Marine Park, 87–88
Reservoirs, 28–29
Resource Assessment Commission (RAC), 93
Rhodes grass, 73
Rum Jungle, 94–95

S

Sandalwood, 57
Saturated soil culture (SSC), 54
Savanna
  aerial photographs, 45
  burning and carbon emission, 44–45
  CAM, 40
  fire management, 42–44
  Melaleuca swamps, 39
  mesotrophic forest, 39
  palaeo-fire records, 42
  re-sprouting capacity, 43
  semi-arid, 43–44
  soil carbon, 40–42
  vegetation, 39–40
Seagrass beds
  cyclone effects, 79
  FRDC, 78
  habitat types, 79
  resources, 78–79
SOC, *see* Soil organic carbon
Soil organic carbon (SOC), 40–41, 56

Solomon Dam, 28
Southern Oscillation climatic cycle, 13
SSC, *see* Saturated soil culture
Stratigraphical records, 29
Sugar cane industry
  C-4 Photosynthesis, 49–50
  elevated CO_2, 50
  ethanol, 50–51
  GCTB, 48–49
Sustainable development
  key factor, 111
  report, 98

T

TDS, *see* Total dissolved solids
Teak forestry, 66–67
Tectonic movements, 1
Total dissolved solids (TDS), 97
Tourism and the reef, 88
Tree litter, 43, 69
Tropical Cyclone Roma, 4
Tully-Murray River systems, 75
Tussock grasses, 44, 69, 101

V

Vapour pressure deficits (VPD), 6
Vascular hydraulic transport system, 12
VPD, *see* Vapour pressure deficits

W

Water-use efficiency (WUE), 13, 17
Weirs/river barrages, 27–28
Wet Tropics World Heritage Area (WTWHA), 1,
          2–3, 22, 106
Wetland
  carbon flux, 31–32
  floodplain, 23, 32
  intermittent rivers, 24–25
  invasive species
    cane toads, 33–34
    Mimosa pigra, 34
    rubberbush, 36
    rubber vine, 35
    Salvinia molesta, 36
    vermin, 36–37
    Ziziphus, 34–35
  underground water, 23–24
Wilderness, 105
Wild Rivers Act, 106, 114–115
WTWHA, *see* Wet Tropics World Heritage Area
WUE, *see* Water-use efficiency